Global Environmental Change

ISSUES IN ENVIRONMENTAL SCIENCE AND TECHNOLOGY

TITLES IN THE SERIES:

1. Mining and its Environmental Impact
2. Waste Incineration and the Environment
3. Waste Treatment and Disposal
4. Volatile Organic Compounds in the Atmosphere
5. Agricultural Chemicals and the Environment
6. Chlorinated Organic Micropollutants
7. Contaminated Land and its Reclamation
8. Air Quality Management
9. Risk Assessment and Risk Management
10. Air Pollution and Health
11. Environmental Impact of Power Generation
12. Endocrine Disrupting Chemicals
13. Chemistry in the Marine Environment
14. Causes and Environmental Implications of Increased UV-B Radiation
15. Food Safety and Food Quality
16. Assessment and Reclamation of Contaminated Land
17. Global Environmental Change

FORTHCOMING:

18. Environmental Impact of Solid Waste Management Activities

How to obtain future titles on publication

A subscription is available for this series. This will bring delivery of each new volume immediately upon publication and also provide you with online access to each title via the Internet. For further information visit www.rsc.org/issues or write to:

Sales and Customer Care Department, Royal Society of Chemistry, Thomas Graham House, Science Park, Milton Road, Cambridge CB4 0WF, UK

Registered Charity Number 207890

Telephone: +44 (0) 1223 432360
Fax: +44 (0) 1223 426017
Email: sales@rsc.org

ISSUES IN ENVIRONMENTAL SCIENCE AND TECHNOLOGY

EDITORS: R. E. HESTER AND R. M. HARRISON

17

Global Environmental Change

ISBN 0-85404-280-6
ISSN 1350-7583

A catalogue record for this book is available from the British Library

Published by The Royal Society of Chemistry, Thomas Graham House,
Science Park, Milton Road, Cambridge CB4 0WF, UK

Registered Charity Number 207890

For further information see our web site at www.rsc.org

Typeset in Great Britain by Vision Typesetting, Manchester
Printed and bound by Bookcraft Ltd, UK

Preface

After much debate based on the accumulating scientific evidence it is now generally, though still not universally, accepted that the Earth's climate is undergoing changes induced by man's activities. Global warming, due primarily to increasing concentrations of 'greenhouse gases' such as carbon dioxide in the Earth's atmosphere, has become a matter of concern well beyond the scientific community, to politicians and to the general public. The consequences of global warming are profound. Climate changes, including major effects on rainfall volumes and patterns, impact immediately on agricultural practice and thus the food supply; rising temperatures of the world's oceans and melting of polar ice will lead to flooding and coastal erosion and may have severely adverse influences on marine ecosystems. In addition to climate change, the industrial activity, which has been so vital to development of nations around the world, has other impacts of global significance, such as air pollution and declining fish stocks. The solution to these problems cannot lie in a reversal of industrial growth and development but must be found in improved and environmentally sensitive practices. All of these aspects of global environmental change are examined in this volume.

The first article, by Sir John Houghton, who is a co-chairman of the Scientific Assessment Working Group of the Intergovernmental Panel on Climate Change (IPCC), reviews the evidence for global warming and outlines the events that led to the setting up of the IPCC in 1988. The breadth of scientific fact and opinion evaluated, and the care taken to articulate carefully and honestly not only the knowledge that is available but also the degree of associated uncertainty, have resulted in the IPCC's reports and statements being powerfully influential with policymakers. The second article, by Johan Kuylenstierna, Kevin Hicks and Mike Chadwick of the Stockholm Environment Institute, University of York, provides an international perspective on the many other problems associated with air pollution. Impacts on human health, crop yield and quality, forestry, buildings and visibility are examined for both developed and developing countries. Air pollution policy in different regions of the globe is reviewed with the aid of case studies, such as the Kathmandu Valley, and prospects for the future are considered.

The influence of climate variability and change on the structure, dynamics and exploitation of marine ecosystems is examined in the third article by Manuel Barange, who is the Director of the GLOBEC International Project Office at the Plymouth Marine Laboratory. As the oceans occupy over 70% of the Earth's surface and absorb twice as much of the sun's radiation as the atmosphere or the land surface, they play a major role in determining climate. The oceans move heat from lower to higher latitudes, thus contributing to the complex variability of

Earth systems. Conversely, the responses of the oceans and their marine biological resources to global warming are complex and imperfectly understood. Phenomena such as the North Atlantic Oscillation and the El Niño-Southern Oscillation and the ecological consequences of industrial fishing are analysed. The fourth article also looks at atmosphere–ocean interactions but with its focus on coastal zones. This is by Robert Nicholls of the Flood Hazard Research Centre at Middlesex University and deals with the vulnerability of human populations to sea-level rises. The fact that the majority of the world's population live within 50 miles of the coast and that coastal populations are growing at twice the global average underlines the importance of the assessment of climate change impacts. In the fifth of the articles Martin Parry and Matthew Livermore of the University of East Anglia review the possible effects of climate change on global agricultural yield potential, on cereal production, food prices and the implications for changes in the number of hungry people in the world. These authors show how modelling may be used to predict the effects of climate change, including quantitative estimates of how the additional numbers of people at risk of hunger depend on the possible mitigation strategies used. The wider issues of human health are the focus of the sixth article, written by Michael Ahern and Tony McMichael of the London School of Hygiene and Tropical Medicine. These include respiratory illnesses arising from air pollution, both indoor and outdoor, microbial diseases carried by water-borne organisms, toxic chemical contaminants in food and the adverse effects of increased ultraviolet radiation associated with stratospheric ozone depletion. Both direct and indirect health effects of climate change are reviewed and consideration is given to the increased risk of conflict which arises when natural resources, such as water, become scarce. The concept of global environmental public goods and services is examined in relation to human health.

The seventh and final article is by Frans Berkhout of the Science and Technology Policy Research Unit (SPRU) at the University of Sussex. This analyses corporate environmental performance and presents some results from a study of European companies which illustrate the growing concerns of businesses with corporate social responsibility. Environmental performance indicators are scrutinized and the relationship between environmental and financial performance is examined with the aid of results from a specific study 'Measuring Environmental Performance of Industry (MEPI)' carried out at SPRU. This article will be of particular interest to industrial managers for its examination of profits and environmental performance and to policymakers for its analysis of policy conclusions drawn from the MEPI study.

The causes and consequences of global environmental change are many and varied. We believe this collection of reviews by a set of distinguished and authoritative contributors provides a balanced overview of this vast subject area. It will be found of particular value by environmental scientists, industrial managers, policymakers and students seeking a readable but factually detailed and reliable review of global warming and other current issues consequent on environmental change.

Ronald E. Hester
Roy M. Harrison

Contents

Issues in Environmental Science and Technology, No. 17
Global Environmental Change

Editors

Ronald E. Hester, BSc, DSc(London), PhD(Cornell), FRSC, CChem

Ronald E. Hester is now Emeritus Professor of Chemistry in the University of York. He was for short periods a research fellow in Cambridge and an assistant professor at Cornell before being appointed to a lectureship in chemistry in York in 1965. He was a full professor in York from 1983 to 2001. His more than 300 publications are mainly in the area of vibrational spectroscopy, latterly focusing on time-resolved studies of photoreaction intermediates and on biomolecular systems in solution. He is active in environmental chemistry and is a founder member and former chairman of the Environment Group of the Royal Society of Chemistry and editor of 'Industry and the Environment in Perspective' (RSC, 1983) and 'Understanding Our Environment' (RSC, 1986). As a member of the Council of the UK Science and Engineering Research Council and several of its sub-committees, panels and boards, he has been heavily involved in national science policy and administration. He was, from 1991 to 93, a member of the UK Department of the Environment Advisory Committee on Hazardous Substances and from 1995 to 2000 was a member of the Publications and Information Board of the Royal Society of Chemistry.

Roy M. Harrison, BSc, PhD, DSc (Birmingham), FRSC, CChem, FRMetS, Hon MFPHM, Hon FFOM

Roy M. Harrison is Queen Elizabeth II Birmingham Centenary Professor of Environmental Health in the University of Birmingham. He was previously Lecturer in Environmental Sciences at the University of Lancaster and Reader and Director of the Institute of Aerosol Science at the University of Essex. His more than 300 publications are mainly in the field of environmental chemistry, although his current work includes studies of human health impacts of atmospheric pollutants as well as research into the chemistry of pollution phenomena. He is a past Chairman of the Environment Group of the Royal Society of Chemistry for whom he has edited 'Pollution: Causes, Effects and Control' (RSC, 1983; Fourth Edition, 2001) and 'Understanding our Environment: An Introduction to Environmental Chemistry and Pollution' (RSC, Third Edition, 1999). He has a close interest in scientific and policy aspects of air pollution, having been Chairman of the Department of Environment Quality of Urban Air Review Group and the DETR Atmospheric Particles Expert Group as well as currently being a member of the DEFRA Expert Panel on Air Quality Standards, the DEFRA Advisory Committee on Hazardous Substances and the Department of Health Committee on the Medical Effects of Air Pollutants.

Contributors

M. J. Ahern, *London School of Hygiene and Tropical Medicine, Keppel Street, London WC1E 7HT, UK*

M. Barange, *GLOBEC IPO, Plymouth Marine Laboratory, Prospect Place, Plymouth PL1 3DH, UK*

F. Berkhout, *SPRU-Science and Technology Policy Research, University of Sussex, Falmer, Brighton BN1 9RF, UK*

M. J. Chadwick, *Stockholm Environment Institute, University of York, York YO10 5DD, UK*

W. K. Hicks, *Stockholm Environment Institute, University of York, York YO10 5DD, UK*

Sir John Houghton, *Intergovernmental Panel on Climate Change, Hadley Centre, Meteorological Office, Bracknell RG12 2SY, UK*

J. C. I. Kuylenstierna, *Stockholm Environment Institute, University of York, York YO10 5DD, UK*

M. T. J. Livermore, *School of Environmental Sciences, University of East Anglia, Norwich NR4 7TJ, UK*

A. J. McMichael, *London School of Hygiene and Tropical Medicine, Keppel Street, London WC1E 7HT, UK*

R. J. Nicholls, *Flood Hazard Research Centre, Middlesex University, Enfield EN3 4SF, UK*

M. L. Parry, *School of Environmental Sciences, University of East Anglia, Norwich NR4 7TJ, UK*

An Overview of the Intergovernmental Panel on Climate Change (IPCC) and Its Process of Science Assessment

SIR JOHN HOUGHTON

1 Background[1]

It has been known for about 175 years that the presence in the atmosphere of 'greenhouse gases' such as carbon dioxide that absorb in the infrared part of the spectrum leads to a warming of the Earth's surface through the 'greenhouse' effect. The first quantitative calculation of the effect on the atmosphere of increased carbon dioxide concentrations was made by the Swedish scientist Svante Arrhenius in 1896. In the 1960s, Charles Keeling and his colleagues began a regular series of accurate observations of atmospheric carbon dioxide concentration from the Mauna Loa Observatory in Hawaii. These showed increasing values as a result of human activities, mainly the burning of fossil fuels. By the 1980s, as the rate of increase of carbon dioxide concentration became larger, the possible impact on the global climate became a matter of concern to politicians as well as scientists. The report of a scientific meeting held at Villach, Austria in 1985 (SCOPE 29, 1986) under the auspices of the Scientific Committee on Problems of the Environment (SCOPE) of the International Council of Scientific Unions (ICSU) began to alert governments and the public at large to the potential seriousness of the issue. Estimates were made that the carbon dioxide concentration could double before the end of the 21st century. In 1986, three international bodies, the World Meteorological Organisation (WMO), the United Nations Environment Programme (UNEP) and the International Council of Scientific Unions (ICSU), who had co-sponsored the Villach conference, formed the Advisory Group on Greenhouse Gases (AGGG), a small international committee with responsibility for assessing the available scientific information about the increase of greenhouse gases in the atmosphere and the likely impact.

[1] A comprehensive account of Global Warming and Climate Change can be found in J. Houghton, *Global Warming: the Complete Briefing*, 2nd Edition, Cambridge University Press, 1997.

Issues in Environmental Science and Technology, No. 17
Global Environmental Change

2 Formation of the IPCC

What was new about the problem of 'global warming' (as the climate change due to the increase of gases began to be called) was that it is an example of *global* pollution or pollution on the *global* scale, *i.e.* pollution emitted by one person locally that has global effects. This can be compared with pollution due to human activities on a *local* scale, of air, water or land that has been around for a very long time. The other example of global pollution which was recognized about the same time is the damage to the ozone layer in the stratosphere that results from the release of small quantities of chemicals containing chlorine (*e.g.* the chlorofluorocarbons or CFCs).

The existence of global pollution requires *global* solutions, *i.e.* solutions that are organized on a global scale. In the late 1980s, therefore, as political concern began to be expressed about the possibility of deleterious climate change, the organization of that concern was international, as indeed had been the work of the scientists on which the political concern was based. In June of 1988 an international conference was staged in Toronto which for the first time pressed for specific international action to mitigate climate change. It was in that year too that world leaders began to speak out about it; for instance, Mrs Thatcher expressed her concern in a speech to the Royal Society of London that was widely publicized.

It was therefore timely that in 1988 a new international scientific body to address the issue, the IPCC, was set up jointly by the World Meteorological Organisation (WMO) and the United Nations Environment Programme (UNEP). Bert Bolin from Sweden, a scientist with a distinguished record of contributions to the science of climate, agreed to chair the IPCC. Three Working Groups were established, WGI to address the science of anthropogenic climate change, WGII to address the impacts and WGIII to address the policy options. I was appointed chairman of WGI and I will illustrate from my experience of that Working Group the work of the IPCC and how, through the IPCC, scientists have been able to assist in the determination of policy.

The establishment of the IPCC followed closely that of the Montreal Protocol which had been set up in the previous year, 1987, by UNEP and WMO to address the problem of the depletion of stratospheric ozone by CFCs and related chlorine-containing chemicals. This problem addressed by the Montreal Protocol was a more limited one than that of global climate change, especially in the range and size of the human activities that contribute to it. However, through the negotiation of the Protocol with its arrangements for inputs from scientists and other experts, methods had begun to be developed in the international arena through which problems of global pollution could be addressed. It was therefore appropriate that the IPCC should build on this experience. The development within the IPCC of ways to involve large numbers of scientists and of formal procedures for peer review in turn influenced the on-going work of the Assessment Panels of the Montreal Protocol.

3 The IPCC 1990 Report[2]

It was agreed at the first meeting of the IPCC that a new assessment of the whole issue of anthropogenic climate change should be prepared. There had, of course, been assessments before of the climate change issue, notably that resulting from the Villach conference (SCOPE 29), again under the chairmanship of Bert Bolin as mentioned in the introduction. The IPCC saw its task as updating previous assessments, but with a difference. Previous assessments had involved relatively few of the world's leading climate scientists. Because of the global nature of the issue that brought with it a large measure of international concern, the IPCC's ambition from the start was to involve as many as possible from the world scientific community in the new assessment.

To assist in the preparation of the WGI report, a small Technical Support Unit was set up within the part of the UK Meteorological Office at Bracknell which was concerned with Climate Research. The report comprised eleven chapters totalling over 300 pages dealing with different components of the scientific issue together with a Policymakers' Summary and an Executive Summary. Twelve international workshops were held to address these different components. One hundred and seventy scientists from 25 countries contributed to the report either through participation in the workshops or through written contributions. A further 200 scientists were involved in the peer review of the draft report. The thorough peer review assisted in achieving a high degree of consensus amongst the authors and reviewers regarding the report's conclusions.

The Policymakers' Summary (20 pages) together with its Executive Summary (2 pages) was based on the conclusions presented in the chapters and was prepared particularly to present to those without a strong background in science a clear statement of the status of scientific knowledge at the time and of the associated uncertainties. In preparing the first draft of the Policymakers' Summary, the Lead Authors of the chapters were first involved; it was then sent out for the same wide peer review as the main report. A revised draft of the Summary was then discussed line by line at a Plenary Meeting of the Working Group attended by government delegates from 35 countries together with Lead Authors from the chapters, and the final wording agreed at that meeting.

A flavour of the style and content of the report is given by the first few paragraphs of the Executive Summary which read as follows:

We are certain of the following:

- there is a natural greenhouse effect which already keeps the Earth warmer than it would otherwise be.
- emissions resulting from human activities are substantially increasing the atmospheric concentrations of the greenhouse gases: carbon dioxide, methane, chlorofluorocarbons (CFCs) and nitrous oxide. These increases will enhance the greenhouse effect, resulting on average in an additional warming of the Earth's surface. The main greenhouse gas, water vapour, will

[2] *Climate Change, the IPCC Scientific Assessment*, eds. J.T. Houghton, G.J. Jenkins and J.J. Ephraums, Cambridge University Press, 1990.

increase in response to global warming and further enhance it.

We calculate with confidence that:

- some gases are potentially more effective than others at changing climate, and their relative effectiveness can be estimated. Carbon dioxide has been responsible for over half the enhanced greenhouse effect in the past, and is likely to remain so in the future.
- atmospheric concentrations of the long-lived gases (carbon dioxide, nitrous oxide and the CFCs) adjust only slowly to changes in emissions. Continued emissions of these gases at present rates would commit us to increased concentrations for centuries ahead. The longer emissions continue to increase at present day rates, the greater reductions would have to be for concentrations to stabilize at a given level.
- the long-lived gases would require immediate reductions in emissions from human activities of over 60% to stabilize their concentrations at today's levels; methane would require a 15–20% reduction.

Based on current model results, we predict:

- under the IPCC Business-as-Usual (Scenario A) emissions of greenhouse gases, a rate of increase of global mean temperature during the next century of about 0.3 °C per decade (with an uncertainty range of 0.2 °C to 0.5 °C per decade); this is greater than that seen over the past 10 000 years. This will result in a likely increase in global mean temperature of about 1 °C above the present value by 2025 and 3 °C before the end of the next century. The rise will not be steady because of the influence of other factors.

Later sections of the summary addressed the scientific uncertainties and the question of the degree to which anthropogenic climate change had been observed in the climate record.

Over the period of the preparation of the IPCC report, a significant change occurred in the attitudes of the scientists involved. To begin with there was a strong feeling, particularly amongst some scientists, that the scientific uncertainty was too large for any useful statement to be made regarding future climate change. However, gradually we all realized our responsibility to articulate carefully and honestly the knowledge which is available, distinguishing clearly between what could be said with a good degree of certainty and the areas where the uncertainty is large. After all, there were many not expert in the science who felt few inhibitions about making forecasts of future climate change – often of an extreme kind. Also we increasingly recognized that there was enough certainty in the science to provide meaningful information regarding the likely future, provided that the uncertainty was also fully explained.

4 Ownership by Scientists and by Governments

Many of the world's leading scientists involved with the understanding of climate and climate change contributed to the report. Inevitably they came mostly from developed countries. However, a significant number of contributors from developing countries were also involved. That so many of the world's scientists contributed or were involved in the review process meant that there was a genuine feeling of ownership of the report by the world scientific community.

The IPCC process led to a significant degree of consensus. It is sometimes pointed out that 'consensus' amongst scientists is not necessarily a sign of scientific health; argument and disagreement are seen to be more usual building blocks of scientific advance. But the 'consensus' achieved by the IPCC is not complete agreement about everything; it is agreement particularly about what we know and what we do not know – distinguishing clearly those matters about which there is reasonable certainty from those where there remains much uncertainty and where there continues to be lively debate and disagreement. It is this limited 'consensus' which is reflected in the Executive Summary of the 1990 IPCC Report which has been widely acclaimed for the clarity and crispness of its presentation.

It was clear from an early stage that not only was the scientific content of the assessment important but also the way in which it was presented. Scientists left to themselves do not always recognize what is relevant to policymakers or present their material with the maximum clarity. Further, the presentation of a scientific document can appear to a policymaker to convey a political message even though none was intended, for instance through the selection of the particular material employed.

It has therefore been helpful in the presentation of the science of climate change to involve policymakers themselves or their representatives in the formulation of the summary of the reports. For instance, they were full participants in the government review process and in the Working Group Plenary Meeting which agreed the wording of the report. The report was greatly improved in its relevance and clarity through their participation. In addition the large number of governments which had a part in the process felt ownership of the report.

The IPCC was therefore able to provide to the Earth Summit at Rio in 1992 a clear assessment of the science of climate change that was owned both by the world scientific community and by governments. These characteristics were essential to providing governments with the confidence to formulate and to sign the Framework Convention on Climate Change at that 1992 Conference and to take appropriate action. They have continued to be essential in the generation of subsequent reports which have provided input to the on-going work of the FCCC, for instance to the Kyoto Protocol of 1997.

5 The Science Policy Interface

The work of the IPCC illustrates the following five important features which I believe should characterize the scientific assessments that form an input to policy making.

The first has already been mentioned, namely the separation of what is known with reasonable certainty from what is unknown or very uncertain. All statements from scientists that have policy implications should make this distinction and should describe and quantify the uncertainty as fully as possible.

Secondly, it has been important for its continued credibility that the IPCC has confined itself in its reports and statements to scientific information and has avoided making judgements or giving advice about policy. Often in the past these areas have been confused. The scientific information must, of course, be comprehensive and must include input from all relevant scientific disciplines, including the social sciences. But in the formulation and presentation of policy options or in making policy judgements the scientific input must be clearly distinguished from the policy judgements and decisions. The importance in environmental decision making of this separation of scientific and other expert assessment from policy judgement is argued in a recent report of the Royal Commission on Environmental Pollution in the UK.[3]

Thirdly, all parts of the assessment process need to be completely open and transparent. IPCC documents including early drafts and review comments have been freely and widely available – adding much to the credibility of the process and its conclusions.

Fourthly, the purpose of an assessment is to take account of all scientific data and all genuine scientific opinion and to elucidate and articulate the best scientific interpretation and conclusions from the information available. Scientific assessments must not start with preconceived assumptions and no compromises must be made to meet any personal or political agendas. A thorough and wide peer review process helps to guarantee the honesty and comprehensiveness of the process.

Fifthly, the scientific information must be integrated in a thoroughly balanced way. The amount of data available concerning climate and climate change is very large and it is easy to select data that fits in with a wide range of preconceived ideas or assumptions. However, by involving so many scientists from the complete range of relevant disciplines it has been possible to develop a balanced integration of the information in a way that has commanded general acceptance.

6 IPCC's Work in Science Applications and in Social Sciences

Working Groups II and III of IPCC have been concerned with the impacts of climate change and adaptation to and mitigation of climate change. Consideration of these has involved not only natural scientists but also experts from many areas of social science, especially economics. As with WGI the aim of the other Working Groups has been to involve scientists with a wide range of expertise and from as many countries as possible. This has been less easy in the social sciences than in the natural sciences where, especially in a subject like the science of climate, there has been a long tradition of scientists working together across national boundaries. However, through the stimulation of the IPCC a substantial international community of social scientists has been brought together to address the variety of problems exposed by the climate change issue.

In this area, as in natural science, the IPCC has stressed the importance of separating the scientific analysis from policy judgements. But it is clearly more difficult when dealing with economic or political analysis to make this separation

[3] A full discussion of the inputs to the environmental policy determination and the process of decision making is given in the 21st Report of the UK Royal Commission on Environmental Pollution, *Setting Environmental Standards*, The Stationery Office, London, 1998.

convincing. Because of this some have argued that the IPCC should not become engaged in analysis in these social science areas. The IPCC has consistently refused to accept that argument, believing that a great deal of useful technical supporting work needs to be done in providing analyses of some of the economic or political options which might be taken up in response to the impact of climate change. Further, it is a great advantage if this work is pursued outside government agencies or other political institutions. Such analyses are essential input for international negotiations regarding the options and essential preparation for decision making.

7 The 1992, 1994 and 1995 IPCC Reports

As soon as the 1990 Report was complete, the IPCC began work on further reports. In 1992, in time for the Earth Summit, a report was produced updating what was known about greenhouse gases,[4] their sources and sinks, and about observations and modelling of climate change. In addition, the 1992 report developed various emission scenarios for greenhouse gases for the 21st century based on a variety of assumptions regarding factors such as world population, economic growth, availability of fossil fuels, *etc.*

The year 1994 saw the production of another report updating the area of radiative forcing.[5] Of particular importance was the new work carried out on the profiles of emissions of carbon dioxide and other greenhouse gases which would lead to the stabilization of these gases in the atmosphere at different levels of concentration.

By 1995 when the IPCC produced its second comprehensive assessment,[6] five years after the first assessment in 1990, the community of scientists involved with the IPCC had become substantially greater. More scientists from more countries were involved both in the report's preparation (about 480 scientists from more than 25 countries) and in its review (over 500 from 40 countries). The participants at the Plenary Meeting of Working Group I that approved the Summary for Policymakers included 177 delegates from 96 countries, representatives from 14 non-governmental organizations and 28 Lead Authors.

Regarding climate change during the 21st century and its likely impacts, the messages of the 1995 report were essentially the same as those of the 1990 report. Some further detail had emerged during the five years in between, especially regarding the likely contribution to climate change from atmospheric aerosol – the small dust particles that are present in the atmosphere as a result of industrial activity. Also, there was more confidence amongst scientists that observed

[4] *Climate Change 1992: the Supplementary Report to the IPCC Assessment*, eds. J. T. Houghton, B. A. Callander and S. K. Varney, Cambridge University Press, 1992.

[5] *Climate Change 1994: Radiative Forcing of Climate Change and an Evaluation of the IPCC IS92 Scenarios*, eds. J. T. Houghton, L. G. Meira Filho, J. Bruce, Hoesung Lee, B. A. Callander, E. Haites, N. Harris and K. Maskell, Cambridge University Press, 1996.

[6] *Climate Change 1995: the Science of Climate Change*, eds. J. T. Houghton, L. G. Meira Filho, N. Harris, A. Kattenberg and K. Maskell, Cambridge University Press, 1996; *Climate Change 1995: Impacts, Adaptation and Mitigation of Climate Change*, eds. R. T. Watson, M. C. Zinowera and R. H. Moss, Cambridge University Press, 1996; *Climate Change 1995: Economic and Social Dimensions of Climate Change*, eds. J. Bruce, Hoesung Lee and E. Haites, Cambridge University Press, 1996.

changes in climate might be the result of anthropogenic change. The WGI Plenary Meeting debated for a considerable time how to express this somewhat greater scientific confidence in the interpretation of the recent climate record, although still surrounded by much uncertainty. A sentence carefully crafted by the meeting was unanimously agreed by the delegates: 'the balance of evidence suggests a discernible human influence on global climate'.

Following the 1995 Report, the IPCC produced several Technical Papers in order to answer particular questions of importance to policymakers. Also several Special Reports were written addressing particular issues. Of especial importance was a report on *Aviation and the Global Atmosphere*[7] carried out in cooperation with the International Civil Aviation Organisation (ICAO); the authors of the report included many from the aviation industry. The report addressed the impact of emissions from aircraft engines on the atmosphere and the climate and what could be done through engine design or through operational procedures to minimize the impact. Of particular interest was the result that the influence of aerosol (small particles) produced from the aircraft emissions together with emissions of water vapour could have a significant effect on the cloud cover at aircraft cruising altitudes, with an effect on climate possibly comparable to that of the carbon dioxide emitted by aircraft.

8 Attacks on the IPCC

The IPCC and its work has been attacked from two fronts; on the one hand by those who argue that the projections of global warming and its consequences have been grossly overplayed and that there is no need for any action, and on the other hand by those who believe that possible serious consequences of global warming have been ignored and that there is urgent need for much more drastic action. Both these points of view have been bolstered by appeals to aspects of the scientific data available and their interpretation.

Most of the controversy regarding these extreme positions has been exposed in the media which, in general, is much more interested in extreme views than in the more balanced approach adopted by the IPCC.[8] Particular points that can be made about the controversy and the attacks that have been directed at the IPCC are as follows.

1. Amongst the scientists who have contributed to the IPCC there is a wide range of views – the uncertainty ranges associated with the IPCC projections of future climate change illustrate this range.
2. All scientists are, of course, interested in genuine scientific debate, the existence of which is fundamental to scientific advance. Contrary to some of the allegations that have been made, the IPCC has not tried to inhibit debate or to minimize the uncertainties that exist.
3. Because, from the wide range of scientific data that exist, it is relatively easy to select data to support any preconceived conclusion or position, the best

[7] *Aviation and the Global Atmosphere*, eds. J. E. Penner, D. H. Lister, D. J. Griggs, D. J. Dokken and M. McFarland, IPCC Special Report published by Cambridge University Press, 1999.

[8] See editorial in *Nature*, 2001, **412**, 103.

place to pursue debate and discussion about the data is in the scientific literature with its discipline of honesty and integrity and tradition of careful peer review. It is regrettable, therefore, that most of those who have challenged the scientific conclusions of the IPCC reports have chosen to pursue that challenge in the media rather than in the scientific literature where a more scientifically useful debate could be conducted.

4. The number of scientists who have taken the extreme positions mentioned above and who have also contributed significantly to climate science is very few compared with the hundreds who have contributed to the IPCC process.
5. Some vicious attacks in the media have been made not so much on the science but on the integrity of the IPCC process and some of the scientists involved in it.[9] That these attacks possess no real foundation is illustrated by the fact that they have received no formal support from any of the delegates to IPCC, even from those countries strongly opposed to any action concerning climate change.

9 The FCCC and the IPCC

The United Nations Framework Convention on Climate Change (FCCC) signed by over 160 countries at the Earth Summit in Rio de Janeiro in June 1992 sets the context in which international discussion regarding appropriate action can be pursued. The development of the Convention's agenda clearly requires continuous scientific and technical input. The Conference of the Parties (COP) to the FCCC has set up a Subsidiary Body for Science and Technological Advice (SUBSTA) to organize this input. The IPCC is working closely with SUBSTA through a Joint Working Group (JWG) to ensure that IPCC assessments are geared to provide the detailed scientific and technical input required.

The Objective of the FCCC is contained in Article 2. It recognizes the need to prevent continued change of the climate and therefore to stabilize the causes of climate change. It reads as follows:

> 'The ultimate objective of this Convention and any related legal instruments that the Conference of the Parties may adopt is to achieve, in accordance with the relevant provisions of the Convention, stabilization of greenhouse gas concentrations in the atmosphere at a level that would prevent dangerous anthropogenic interference with the climate system. Such a level should be achieved within a time frame sufficient to allow ecosystems to adapt naturally to climate change, to ensure that food production is not threatened and to enable economic development to proceed in a sustainable manner.'

The IPCC has been at pains to explain that what constitutes 'dangerous' is a policy not a scientific decision. But the need to make such policy decisions immediately raises many scientific and technical questions. For instance, what carbon dioxide emission profiles will lead to stabilization of atmospheric concentration and by when? What effect will current proposed emission

[9] See, for instance, 'Open Letter to Ben Santer', *Bull. Am. Met. Soc.*, 1996, **77**, 1962–1967.

limitations by developed countries have on atmospheric concentrations? What technologies, policies and measures might be available for mitigating climate change? How vulnerable are different regions of the world to possible climate change? IPCC Technical Papers[10] have addressed the detail of some of the issues involved in answering these questions.

An issue which has been highlighted by the third session of the COP at Kyoto in 1997 is that of the contributions which are made by deforestation, aforestation, reforestation and changes in land use to the sources or sinks of greenhouse gases, especially of carbon dioxide. This is an area where what is meant by different human activities (*e.g.* de-, a-, or re-forestation) requires very careful definition, where there is much scientific uncertainty and where there are significant possibilities for the propagation of perverse incentives. The IPCC is already very involved in this area through its work on the development of detailed guidelines[11] for the production of national inventories of greenhouse gases which include both sources and sinks. It is these IPCC Guidelines to which the Kyoto Protocol refers. The IPCC has also developed further analysis and assessment[12] regarding the interpretation of the Protocol and the use of the Guidelines.

An important basic principle that underlies any appropriate response to the possibility of climate change is the Precautionary Principle. This is stated clearly in Article 3 of the FCCC where the Parties to the FCCC are instructed to

'take precautionary measures to anticipate, prevent or minimize the causes of climate change and mitigate its adverse effects. Where there are threats of serious or irreversible damage, lack of full scientific certainty should not be used as a reason for postponing such measures, taking into account that policies and measures to deal with climate change should be cost-effective so as to ensure global benefits at the lowest possible cost.'

The FCCC, in this statement of the Precautionary Principle and in its Objective, places action concerning climate change clearly in the context of sustainable development. The balance that this implies between environmental protection on the one hand and economic development on the other must be based on the best possible scientific, economic and technical analyses of all the factors involved. This requirement provides a strong impetus for the wide range and the high quality of the work that goes into the IPCC assessments.

[10] *Technologies, Policies and Measures for Mitigating Climate Change* (Tech Paper No 1, 1996); *An Introduction to Simple Climate Models Used in the IPCC Second Assessment Report* (Tech Paper No 2, 1997); *Stabilization of Atmnospheric Greenhouse Gases: Physical, Biological and Socio-economic Implications* (Tech Paper No 3, 1997); *Implications of Proposed CO_2 Emissions Limitations* (Tech Paper No 4, 1997).

[11] *IPCC Guidelines for National Greenhouse Gas Inventories*, IPCC Secretariat, World Meteorological Organisation, Geneva.

[12] *Land Use, Land-Use Change and Forestry*, ed. R.T. Watson, I.R. Noble, B. Bolin, N.H. Ravindranath, D.J. Verardo and D.J. Dokken. IPCC Special Report, published by Cambridge University Press, 2000.

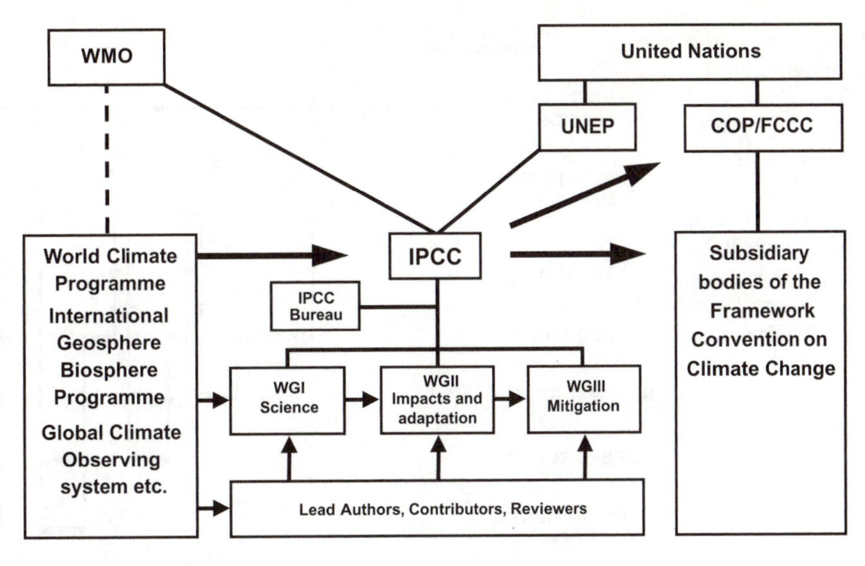

Figure 1 Structure of the IPCC 1997–2001 showing the Working Groups (WGs), the parent organizations of the IPCC, the World Meteorological Organisation (WMO) and the United Nations Environment Programme (UNEP), and showing the IPCC's links to International Scientific Organizations and to the Conference of the Parties (COP) of the Framework Convention on Climate Change (FCCC)

10 The Third Assessment Report

In 2001 the IPCC published its Third Assessment Report (TAR) that addressed in a comprehensive manner the whole field of climate change: the scientific basis (Working Group I), the likely impacts on human communities and on ecosystems and how adaptation might be achieved (Working Group II) and the options for mitigation and its likely cost (Working Group III). In order to explain more fully the procedures now followed by the IPCC, the preparation of the TAR will be described in some detail, after which a brief summary will be provided of some its main conclusions.

In Figure 1 is shown the current structure of the IPCC and its connections with other international bodies. Note that the IPCC is not a body that itself carries out research; in its assessments it relies heavily on the work of international research organizations, for instance the World Climate Research Programme (WCRP) and the International Geosphere Biosphere Programme (IGBP). Each of the Working Groups is led by a Bureau consisting of two co-chairs (one from a developed and one from a developing country) and six vice-chairs, each of whom comes a different geographical region (according to the six geographical regions as defined by the WMO).

The key steps, covering a period of over two and a half years, in the preparation of the IPCC TAR WGI report are illustrated in Figure 2. Similar timetables were followed by the other Working Groups. At the initial scoping meeting in the summer of 1998, over a hundred scientists met for a preliminary discussion about the content and scope of the new report. The headings and outlines for 14 chapters were formulated. In addition, there would be a Technical Summary (TS) about 50 pages long and a Summary for Policymakers (SPM) about 15 pages long. Following that meeting, nominations for Lead Authors were requested very widely from scientific organizations and from governments. From the large number of nominations received, Lead Authors were selected by the WGI Bureau

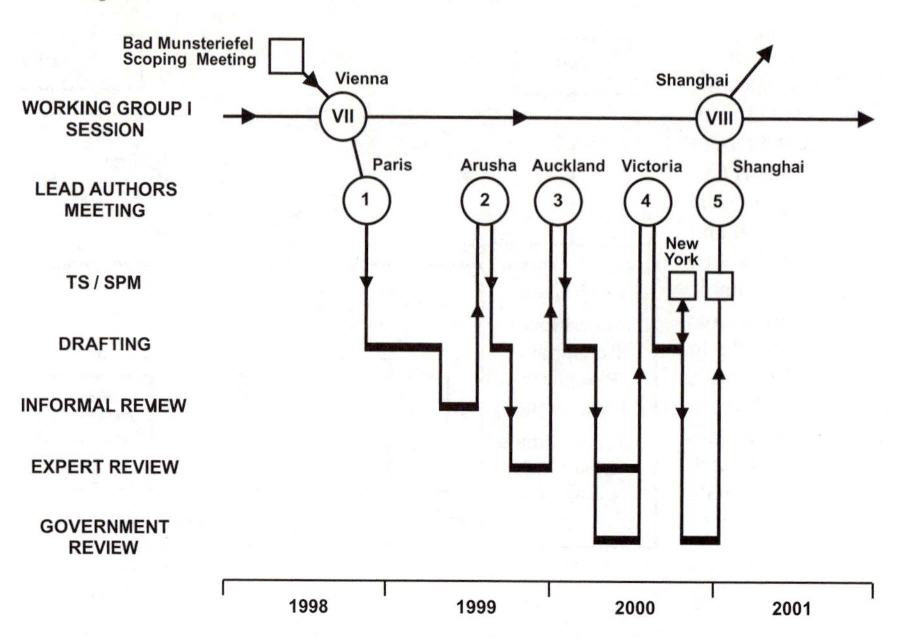

Figure 2 Key steps in the preparation of the Working Group I component of the IPCC Third Assessment Report (adapted from Figure 4 of J. Zillmann, in *Focus*, May–June 2001, Australian Academy of Technological Sciences and Engineering)

for the 14 chapters, representing a range of disciplines and countries; for each chapter at least one Lead Author came from a developing country. For each chapter, one or two Convening Lead Authors were appointed along with between 4 and 12 Lead Authors, depending on the length of the chapter and the breadth of its coverage. Contributions for consideration by the Lead Authors were also invited widely from the scientific community; those who contributed substantially (typically numbering 50 per chapter) are listed alongside the Lead Authors as Contributing Authors. The chapter drafting process began at a Lead Authors' meeting in Paris in late 1998.

The review process is fundamental to the IPCC and the credibility of its reports. For the TAR there were four separate stages of review between mid-1999 and early 2001. First, a draft known as the zero draft was sent in June 1999 to a few expert scientists for informal review. The results of the review were used by the Lead Authors meeting in Arusha, Tanzania, to assist in revising the chapters to produce the first draft that was then sent out for formal review to over 300 experts. These had been nominated by countries, by the Lead Authors, or were scientists who had offered to contribute to the review process. A very large number of review comments were received. All were logged and formally responded to by the Lead Authors at a meeting in Auckland in February 2000.

The Auckland meeting produced a second draft of the chapters that was then sent to governments for review and also sent again to the experts who had responded to the expert review. Governments typically asked scientists within their countries to assist in the review, the scientists' comments being consolidated by the governments before the submission of their review. All these comments were logged and used to produce the final version of the chapters at a Lead Authors' meeting in Victoria, Canada, in July 2000.

Because of some controversy that was raised about the role of Lead Authors in the final editing of one of the chapters of the SAR, a new procedure was introduced for the TAR. Two Review Editors were appointed for each of the chapters, one being a member of the WGI Bureau and the other an independent scientist who was not otherwise involved in the chapter preparation. The role of the Review Editors was to watch over the reviewing of the expert and government comments and to certify that all the comments were appropriately considered and responded to by the Lead Authors. A particular responsibility was to ensure that genuine differences of view among experts were adequately reflected in the revised text. The Lead Authors have welcomed the presence and assistance of the Review Editors and recognized their role in enhancing the credibility of the process and in protecting the Lead Authors from inappropriate or unfounded criticism.

For the drafting of the Technical Summary (TS) a team of Lead Authors with strong representation from the chapters was set up by the WGI Bureau; four members of the Bureau also agreed to act as Review Editors for the TS. A draft of the TS was agreed at the Auckland meeting in February 2000 and sent out for government and expert review along with the chapters. For the Summary for Policymakers (SPM), the Bureau set up a drafting team of about 60, consisting of all the Bureau members, a good representation of Lead Authors and a few others with experience of industry and policy or to ensure adequate geographical representation. A draft of the SPM was also produced at the Auckland meeting and sent out to governments and experts for review.

Following detailed review of both the TS and the SPM at the Victoria Lead Authors' meeting in July 2000, a special meeting of the TS Lead Authors and the SPM writing team was held in New York in October 2000 to rewrite both documents in the light of the final drafts of the chapters as revised at and subsequent to the Victoria meeting. The revised chapters, TS and SPM were then sent to governments in October 2000 in preparation for the Plenary meeting of WGI held in January 2001.

A new feature in the TAR compared with previous reports is the treatment of uncertainty. During the preparation of the TAR a great deal of thought was given to the quantification of uncertainty. In some cases, data are available that enable probabilities of certainty to be estimated by well defined statistical procedures. In other cases scientific judgement plays a large part in the estimation of uncertainty. In either case, what is most helpful to policymakers is that words used to describe levels of uncertainty are employed in a consistent and, so far as is possible, a quantified manner. In the SPM and the TS of WGI therefore, words defined as follows were used where appropriate to indicate estimates of confidence: *virtually certain* (greater than 99% chance that a result is true); *very likely* (90–99% chance); *likely* (66–90% chance); *medium likelihood* (33–66% chance); *unlikely* (10–33% chance); *very unlikely* (1–10% chance); *exceptionally unlikely* (<1% chance). Slightly different definitions were employed by WGII.*

The purpose of the WGI Plenary meeting held in Shanghai in January 2001 was to debate the SPM and agree its wording in detail. The meeting had to be satisfied that the SPM presented comprehensive, accurate and balanced science

* WGII used: *very high* (95% or greater), *high* (67–95%), *medium* (33–67%), *low* (5–33%) and *very low* (5% or less).

in a way that was relevant to and understandable by policymakers. Many scientific questions were raised, but also a large proportion of the meeting was devoted to the detailed presentation of the report's material. In addition the meeting had to formally accept the TS and the underlying chapters as documents that had been through the expert and government review processes, that presented a comprehensive, objective and balanced view of the science of climate change and that were consistent with the SPM.

The basic documentation for the 99 government delegations that attended the Plenary consisted of the SPM, TS and 14 chapters as sent out in October 2000 together with about 100 pages of government and expert review comments on the SPM and a further draft of the SPM which had been revised by the SPM drafting team in the light of the review comments immediately prior to the meeting.

To ensure the integrity of the science as the SPM was debated, 45 scientists representing the Lead Authors of the chapters were present at the Plenary. A good rapport developed between the delegates and the scientists. Many questions were raised by delegates regarding the science, in particular about the degree of certainty about various critical statements and its quantification. Questions were also raised about the presentation and its clarity, in particular how it might translate from the agreed English text into other languages. A number of changes were made to avoid possible misunderstandings. As a result, the SPM was significantly improved from the original draft, both in its accuracy and clarity. A substantial number of changes, mostly of an editorial nature, were made in the underlying chapters to ensure consistency with the finally agreed SPM. Immediately after the meeting in Shanghai, the SPM was put on to the IPCC web site (www.ipcc.ch). Within one month, over one million hits on the site were recorded, demonstrating the large interest throughout the world in the IPCC report.

It is sometimes suggested that because it is an intergovernmental meeting, the SPM that is finally approved cannot be free of political bias.[13] But there are three particular features of the process that work strongly to a bias-free result. First, the 99 delegations present at Shanghai represent the complete range of the political spectrum of view regarding climate change. Secondly, I have already mentioned the active involvement of 45 scientists representing the underlying chapters. Thirdly, although there were very lively and, in some cases, long debates, there was no dissent from either delegations or scientists regarding the final agreed text. It can be said therefore with confidence that no wording was included or added and no changes were made for political or ideological reasons.

Allegations have sometimes been made that the Summaries for Policymakers of IPCC Reports are inconsistent either in content or balance with the underlying chapters. A 10 page summary of a 1000 page document cannot, of course, reflect all the detail or nuances present in the longer document. The National Academy of Sciences of the USA were recently asked to review the WGI TAR; they confirmed its scientific integrity and the scientific consistency between the SPM and the underlying chapters.[14]

The three volumes of the IPCC TAR, one for each Working Group, were published by Cambridge University Press in July 2001. Each is approximately

[13] See editorial and news feature article in *Nature*, 2001, **412**, 103, 112–114.

[14] Report of the US National Academy of Sciences on the IPCC Third Assessment Report, June 2001.

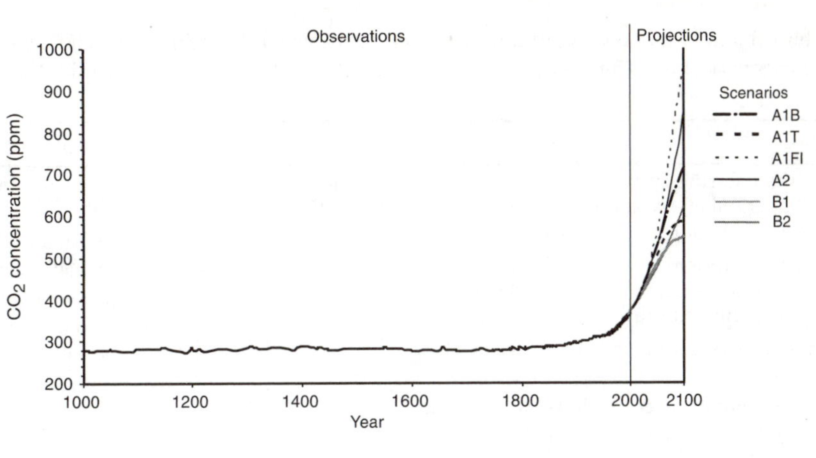

Figure 3 Atmospheric carbon dioxide concentration from 1000 to 2000 from ice core and firn data supplemented with data from direct atmospheric samples over the past few decades. From 2000 to 2100 are shown projections of carbon dioxide concentrations based on six different scenarios of carbon dioxide emissions [from *summary for Policymakers (SPM) of Climate Change* 2001, Synthesis Report, part of the Third Assessment Report (TAR) of the IPCC, published by Cambridge University Press 2001. Details of the scenarios are in the IPCC Special Report on Emission Scenarios (SRES), IPCC 2000]

1000 pages and contains many thousands of references to the scientific literature.

A new feature of the TAR is the Synthesis Report that puts together material of particular policy relevance from the three Working Group reports. It is structured as the answers to nine policy related questions within a relatively short document of about 40 pages that provides extensive references to the Working Group reports. In addition there is a Summary for Policymakers agreed in the same manner as the SPMs of the Working Groups at a meeting of the IPCC Plenary held in London in September 2001.

To conclude this section some of the main findings of the TAR are briefly presented. Table 1 taken from the SPM of the Synthesis Report summarizes some of the most robust findings of the three parts of the report together with the main uncertainties.

Three Figures illustrate some of the most important of the findings associated with the basic science. Figure 3 illustrates the carbon dioxide concentration in the atmosphere as observed over the last millennium, showing the increase since 1850 due largely to the burning of fossil fuels that has taken the concentration to its highest value for at least 420 000 years. Also shown is the projected growth over the next century under a wide range of possible scenarios of carbon dioxide emissions. Figure 4 shows the surface temperature over the northern hemisphere from the year 1000 AD and the global average temperature over the last 140 years and as projected to the year 2100 from climate models for a wide range of greenhouse gas emission scenarios. The 1990s are very likely to have been the warmest decade globally for the period of the instrumental record and for the northern hemisphere likely to have been the warmest decade over the last millennium. The rate of increase projected for the 21st century is very likely to be without precedent over the last 10 000 years, based on paleoclimate data.

Figure 5 illustrates the global averaged temperature over the 20th century as observed and as reconstructed from climate models. The range of results from different models illustrates a significant degree of unforced climate variability that is also reflected in the observations. The models include the forcing effects of increased greenhouse gases (dominant over the last 30 years), increases in sulfate and other aerosols, volcanoes and possible variations in the radiation from the

Table 1 Robust findings and key uncertainties[a] [from *Summary for Policymakers (SPM) of Climate Change 2001*, Synthesis Report, part of the Third Assessment Report (TAR) of the IPCC, published by Cambridge University Press 2001]

Robust findings		*Key uncertainties*
Observations show Earth's surface is warming. Globally, 1990s very likely warmest decade in instrumental record (Figure 4) Atmospheric concentrations of main anthropogenic greenhouse gases [CO_2 (Figure 3), CH_4, N_2O, and tropospheric O_3] increased substantially since 1750 Some greenhouse gases have long lifetimes (*e.g.* CO_2, N_2O and PFCs) Most of observed warming over last 50 years likely due to increases in greenhouse gas concentrations due to human activities (Figure 5)	*Climate change and attribution*	Magnitude and character of climate variability Climate forcings due to anthropogenic aerosols (particularly indirect effects) Relating regional trends to anthropogenic climate change
CO_2 concentrations increasing over 21st century virtually certain to be mainly due to fossil-fuel emissions (Figure 3) Stabilization of atmospheric CO_2 concentrations at 450, 650 or 1000 ppm would require global anthropogenic CO_2 emissions to drop below year 1990 levels, within a few decades, about a century, or about 2 centuries, respectively, and continue to decrease steadily thereafter to a small fraction of current emissions. Emissions would peak in about 1–2 decades (450 ppm) and roughly a century (1000 ppm) from the present For most SRES scenarios, SO_2 emissions (precursor for sulfate aerosols) are lower in the year 2100 compared with year 2000	*Future emissions and concentrations of greenhouse gases and aerosols based on models and projections with the SRES† and stabilization scenarios†*	Assumptions underlying the wide range[b] of SRES emissions scenarios relating to economic growth, technological progress, population growth, and governance structures (lead to largest uncertainties in projections). Inadequate emission scenarios for ozone and aerosol precursors Factors in modelling of carbon cycle including effects of climate feedbacks[b]
Global average surface temperature during 21st century rising at rates very likely without precedent during last 10 000 years (Figure 4) Nearly all land areas very likely to warm more than global average, with more hot days and heat waves and fewer cold days and cold waves Rise in sea level during 21st century that will continue for further centuries Hydrological cycle more intense. Increase in globally averaged precipitation and more intense precipitation events very likely over many areas Increased summer drying and associated risk of drought likely over most mid-latitude continental interiors	*Future changes in global and regional climate based on model projections with SRES scenarios*	Assumptions associated with a wide range[c] of SRES scenarios, as above Factors associated with model projections,[c] in particular climate sensitivity, climate forcing, and feedback processes, especially those involving water vapour, clouds, and aerosols (including aerosol indirect effects) Understanding the probability distribution associated with temperature and sea-level projections The mechanisms, quantification, time scales, and likelihoods associated with large-scale abrupt/non-linear changes (*e.g.* ocean thermohaline circulation) Capabilities of models on regional scales (especially regarding precipitation) leading to inconsistencies in model projections and difficulties in quantification on local and regional scales

Table 1 *(cont.)*

Robust findings		*Key uncertainties*
Projected climate change will have beneficial and adverse effects on both environmental and socio-economic systems, but the larger the changes and the rate of change in climate, the more the adverse effects predominate The adverse impacts of climate change are expected to fall disproportionately upon developing countries and the poor persons within countries Ecosystems and species are vulnerable to climate change and other stresses (as illustrated by observed impacts of recent regional temperature changes) and some will be irreversibly damaged or lost In some mid- to high latitudes, plant productivity (trees and some agricultural crops) would increase with small increases in temperature. Plant productivity would decrease in most regions of the world for warming beyond a few °C Many physical systems are vulnerable to climate change (*e.g.* the impact of coastal storm surges will be exacerbated by sea-level rise, and glaciers and permafrost will continue to retreat)	*Regional and global impacts of changes in mean climate and extremes*	Reliability of local or regional detail in projections of climate change, especially climate extremes Assessing and predicting response of ecological, social (*e.g.* impact of vector- and water-borne diseases), and economic systems to the combined effect of climate change and other stresses such as land-use change, local pollution, *etc.* Identification, quantification, and valuation of damages associated with climate change
Greenhouse gas emission reduction (mitigation) actions would lessen the pressures on natural and human systems from climate change Mitigation has costs that vary between regions and sectors. Substantial technological and other opportunities exist for lowering these costs. Efficient emissions trading also reduces costs for those participating in the trading Emissions constraints on Annex I countries have well-established, albeit varied, 'spill-over' effects on non-Annex I countries National mitigation responses to climate change can be more effective if deployed as a portfolio of policies to limit or reduce net greenhouse gas emissions Adaptation has the potential to reduce adverse effects of climate change and can often produce immediate ancillary benefits, but will not prevent all damages Adaptation can complement mitigation in a cost-effective strategy to reduce climate change risks; together they can contribute to sustainable development objectives Inertia in the interacting climate, ecological, and socio-economic systems is a major reason why anticipatory adaptation and mitigation actions are beneficial	*Costs and benefits of mitigation and adaptation options*	Understanding the interactions between climate change and other environmental issues and the related socio-economic implications The future price of energy, and the cost and availability of low-emissions technology Identification of means to remove barriers that impede adoption of low-emission technologies, and estimation of the costs of overcoming such barriers Quantification of costs of unplanned and unexpected mitigation actions with sudden short-term effects Quantification of mitigation cost estimates generated by different approaches (*e.g.* bottom-up *vs.* top-down), including ancillary benefits, technological change, and effects on sectors and regions Quantification of adaptation costs

[a]In this report a *robust finding* for climate change is defined as one that holds under a variety of approaches, methods, models, and assumptions and one that is expected to be relatively unaffected by uncertainties. *Key uncertainties* in this context are those that, if reduced, may lead to new and robust findings in relation to the questions of this report. This table provides examples and is not an exhaustive list.
[b]Accounting for these above uncertainties leads to a range of CO_2 concentrations in the year 2100 between about 490 and 1260 ppm
[c]Accounting for these above uncertainties leads to a range for globally averaged surface temperature increase, 1990–2100, of 1.4–5.8 °C (Figure 4) and of globally averaged sea-level rise of 0.09–0.88 m.
†See N. Nakicenovic *et al.*, *IPCC Special Report on Scenarios*, Cambridge University Press, 2000.

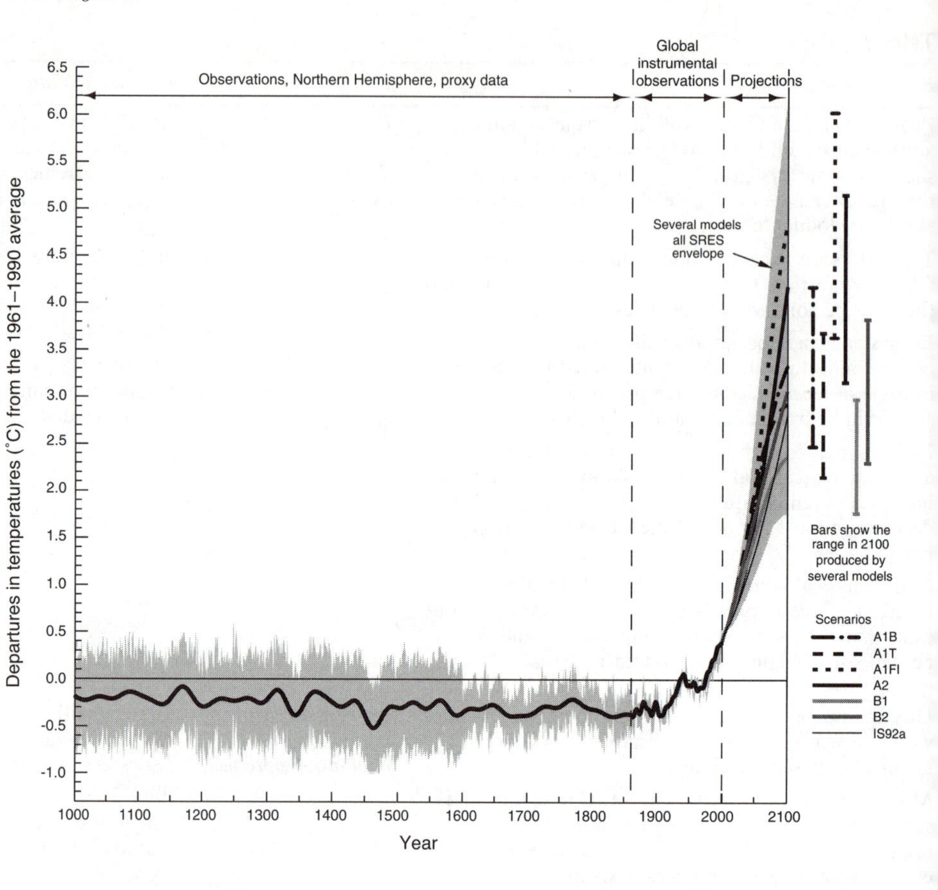

Figure 4 From 1000 to 1860 are shown variations in the average surface temperature of the Northern Hemisphere (adequate data from the Southern Hemisphere not available) constructed from proxy data (tree rings, corals, ice cores and historical data). The line shows the 50 yr average, the grey region the 95% confidence limit in the annual data. From 1860 to 2000 are shown variations in observations of globally and annually averaged surface temperature from the instrumental record; the line shows the decadal average. From 2000 to 2100 are shown projections of globally averaged surface temperature for different scenarios of greenhouse gas emissions as estimated by a model with average climate sensitivity. The grey region marked 'several models, all scenarios envelope' shows in addition the range of results when different models are used with a range of climate sensitivities [*from Summary for Policymakers* (*SPM*) *of Climate Change* 2001, Synthesis Report, part of the Third Assessment Report (TAR) of the IPCC, published by Cambridge University Press 2001. Details of scenarios are in the IPCC Special Report on Emission Scenarios (SRES), IPCC, 2000.

sun. All these influences need to be included to achieve the agreement shown. That such agreement between models and observations can be achieved demonstrates the substantial improvement in the capability of models that has been achieved during the five years since the IPCC SAR.

11 The Future Work of the IPCC

It is a little more than ten years since there was wide realization of the potential danger of anthropogenic climate change to the world community and since it became a significant political issue. During this relatively short time substantial progress has been made. The world's scientists have carefully articulated what is known about the likely climate change (together with the nature of the substantial uncertainty regarding it) and the governments of the world have signed up to the Framework Convention on Climate Change and agreed the first steps towards mitigation measures. Further progress will be dependent on there being full understanding of the issues by all sections of the community in all countries. It is, for instance, important that the general public are adequately informed about all aspects of the environmental problems under assessment so that, so far as is possible, they too can be part of the decision making process.

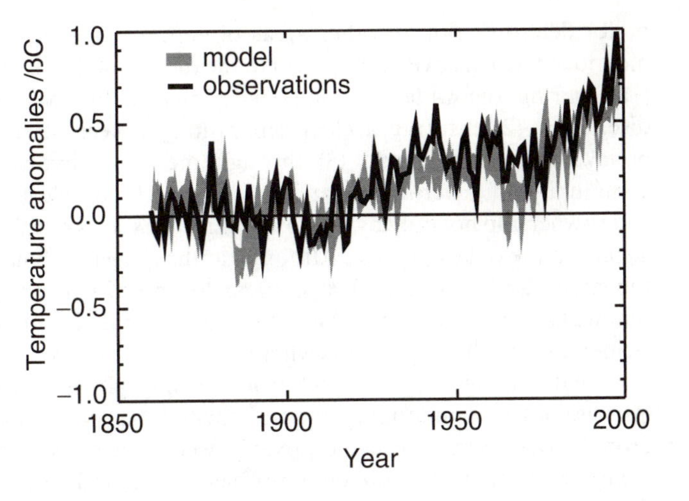

Figure 5 Surface temperature variations (°C) from an ensemble of climate model simulations (grey band) are compared with observations (full line) for the period of the instrumental record, 1860–2000 [from *Summary for Policymakers (SPM) of Climate Change 2001*, Synthesis Report, part of the Third Assessment Report (TAR) of the IPCC, published by Cambridge University Press 2001]

More could be done through the media and otherwise to bring this about.

Of particular importance is that the world industrial sector – which after all is the engine of change – should be adequately informed about environmental issues. As with scientists and governments, industry also needs to feel ownership of the expert assessments in the way that was achieved for the IPCC report on Aviation and Global Climate Change. Leading industrialists are also beginning to see matters of environmental concern as presenting opportunities rather than threats. For instance, in a speech on global climate change, John Browne, the Chief Executive of the British Petroleum Company said.[15]

'No single company or country can solve the problem of climate change. It would be foolish and arrogant to pretend otherwise. But I hope we can make a difference – not least to the tone of the debate – by showing what is possible through constructive action'.

Guidelines for future IPCC work are therefore:

- to continue with an open, transparent, rigorous assessment process involving as many in the world scientific community as possible;
- to ensure the relevance of the assessment to the policy needs of the FCCC;
- to increasingly seek contributions from those in industry, business and commerce so as to stimulate the active participation of the industrial sector in the action required for adaptation to and for mitigation of deleterious climate change;
- to integrate the scientific, technical and economic analyses in ways that illuminate as clearly as possible the various policy options.

12 The IPCC in the Context of Other Global Issues

The question is often asked as to whether the IPCC provides a pattern or a model for providing the means for scientists from all disciplines to provide input to

[15] From a speech in Berlin, 30 September 1997.

policy determination in other areas of concern. The elements which we have mentioned which have been critical to the success of the IPCC have been those of: (1) ensuring the widest possible participation by experts from all relevant disciplines; (2) ensuring a clear separation between scientific assessment and policy determination and (3) through the close involvement of the expert community, of governments and, more recently, of relevant industry, ensuring wide ownership not only by the community of experts but also by those who have a stake in the policy process. Although anthropogenic climate change is perhaps the largest and most complex problem concerned with the global environment that we face, it is not unique. Models similar to that of the IPCC with its essential elements could be applied elsewhere.

Climate change is, of course, not the only global issue facing the planet. Because of the commonality between them, other problems such as population growth, over resource use and poverty need to be considered alongside climate change and other environmental problems (*e.g.* biodiversity loss, deforestation, desertification, *etc.*). Addressing all these together is a tremendous challenge and there is a clear imperative for more mutual understanding and more genuine cooperation in the international arena. The development of the work of the IPCC has demonstrated the enormous capacity for the international community to work together towards the common aims of care for humanity and care for the environment and provides encouragement in the belief that problems as complex as that of climate change are capable of solution.

13 Acknowledgements

The achievements of the IPCC have only been possible because of the dedicated work of many people, contributors, authors, reviewers and supporting staff at the Technical Support Units of the Working Groups and the IPCC Secretariat. For Working Group I, dealing with the scientific basis, I particularly wish to acknowledge the Lead Authors of the chapters and the Technical Summaries and the members of the Technical Support Unit in Bracknell led in turn by Geoffrey Jenkins (for the 1990 Report), Bruce Callander (for the 1995 Report) and David Griggs (for the 2001 Report) who have played a absolutely key part in organizing the work of the hundreds of scientists worldwide who have been involved.

A Perspective on Global Air Pollution Problems

JOHAN C. I. KUYLENSTIERNA, W. KEVIN HICKS AND MICHAEL J. CHADWICK

1 Introduction

Few of Man's activities have zero environmental effect. For centuries, societies have suffered a wide variety of environmental damage, not least in terms of the air pollution that primitive, uncontrolled methods of combustion caused. People of Roman Britain, with fires in their huts, found the smoke a hazard to health and examination of human skulls shows a pitting that indicates a high frequency of sinusitis. The UK was the first country to widely experience massive industrial development based on coal. Between 1580 and 1680, imports of coal to London increased from 20 000 tonnes a year to 360 000 tonnes and London became a city of 'great stinking fogs'.[1] These were only really to disappear some years after the episode in December 1952 when cold temperatures and moist air resulted in the condensation of moisture on particles in the air from coal combustion. Over a radius of 20 miles the resulting fog rendered the atmosphere opaque and it was found that in the week ending 13 December 1952 there had been an excess of 2851 over the usual weekly number of deaths.[1] In response, the *Clean Air Acts* of 1956 and 1968 dealt mainly with smoke in cities. Two decades after the smog disaster the *Control of Pollution Act* (1974) was brought in and systematically began the long haul towards comprehensive improvements in air quality as part of overall environmental protection. Similar experiences were encountered across Europe and North America during the 20th century.

Industrialization, Urbanization and Intensity of Energy Use

The steady advance, world-wide, of industrialization, urbanization and intensification (the amount of energy or materials use per unit of product delivered) has had significant effects on the environment, not least on the pollution of the atmosphere. Figure 1 shows the remarkable increase in energy use over the previous 100 years and how this has been tracked by the rise in SO_2 emissions. Figure 2 indicates how trends in efficiency of the use of raw materials, including fuel, proceed only slowly and sequentially. Reduction in the intensity of energy use by significant amounts (the lowering of the energy requirement to manufacture

[1] P. Brimblecombe, *The Big Smoke*, Methuen, London, 1987.

Issues in Environmental Science and Technology, No. 17
Global Environmental Change

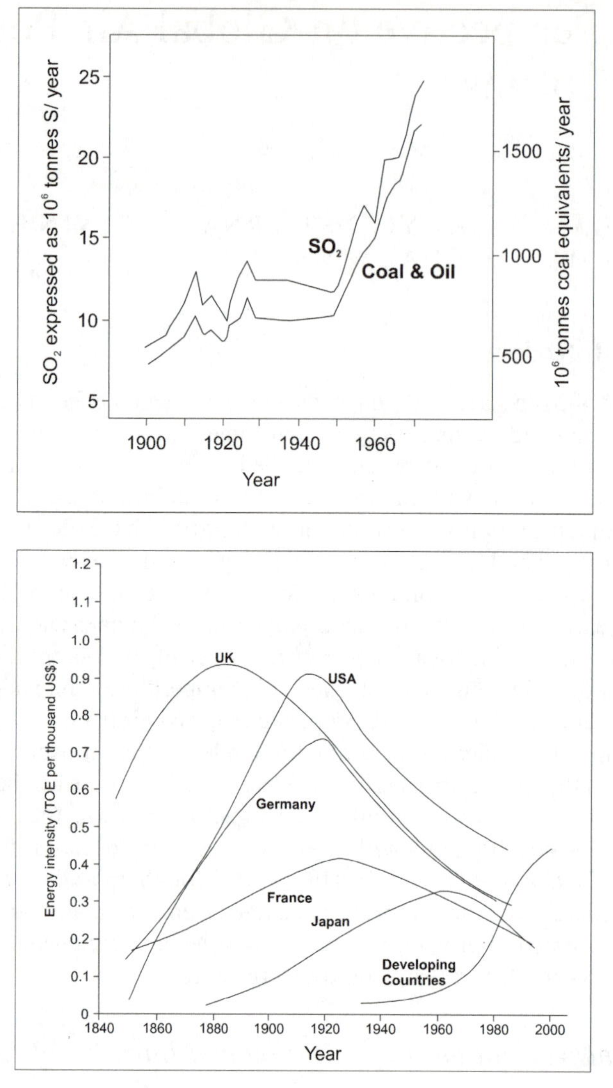

Figure 1 The increase since between 1900 and 1970 of SO_2 emissions in Europe mirroring energy use[2]

Figure 2 Long-term historical evolution of the energy intensity (smoothed curves) of five industrialized countries and the projected trend for developing countries (redrawn in part from ref. 3)

a unit product) will take many years to achieve in some of the less-developed countries.

Defensive Expenditures

In highly-developed industrialized regions of the World, such as Europe, North America and Japan, one aspect of economic development that gives a good indication of the serious nature of the pollution that has followed in its wake is the annual cost incurred on 'defensive expenditure'. These defensive expenditures

[2] B. Fjeld, *Forbruk av fossilt brennsle i Europa og Utslipp av SO_2 i perioden 1900–1972*, NILU Teknisk Notat 1/76, 1976.

[3] J. M. Martin, 1988. L'intensité énergétique de l'activité économique: les évolutions de très longues périodes livrent-elles des enseignements utiles, *Economie et Sociétés*, 1988, **49**, 27.

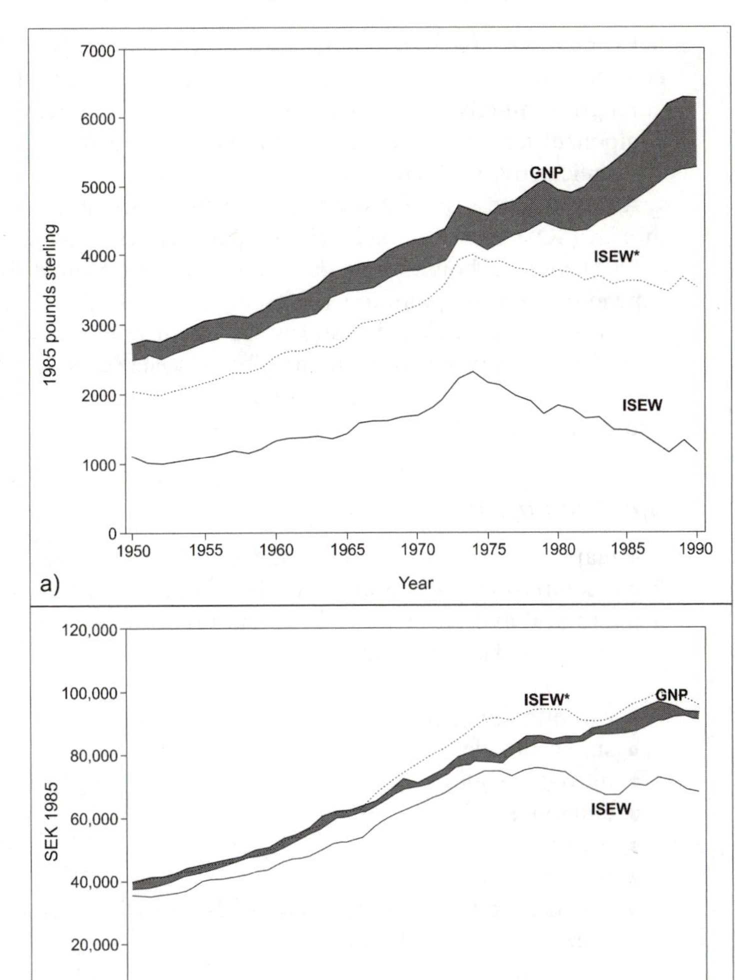

Figure 3 The Index of Sustainable Economic Welfare (ISEW), ISEW* and Gross National Product (GNP) per capita in (a) the UK and (b) Sweden between 1950 and 1990. The shaded area represents per capita costs of defensive expenditure. The higher ISEW* does not include the costs of depletion of non-renewable resources and environmental damage; (a) redrawn from ref. 4; (b) redrawn from ref. 5

represent the costs of positive action taken to protect society from some or all of the adverse effects of economic development. These may be used to 'adjust' Gross National Product (GNP) as an indicator of development (and welfare) to take account of elements of environmental degradation, such as pollution. In fact, GNP includes defensive expenditures (such as the costs of air pollution control) which thus increase the GNP value. Measures of social and economic welfare should rather subtract expenditure necessary to defend the population from the unwanted side-effects of production from GNP.[6]

[4] T. Jackson and N. Marks, *Measuring Sustainable Economic Welfare – a Pilot Index 1950–1990*, Stockholm Environment Institute, Stockholm and New Economics Foundation, London, 1994.

[5] T. Jackson and S. Stymne, *Sustainable Economic Welfare in Sweden – 1950–1992*. Stockholm Environment Institute, Stockholm, 1996.

[6] H. Daly and J. Cobb, *For the Common Good – Redirecting the Economy Towards Community, Environment and a Sustainable Future*, Beacon, Boston, MA, 1989.

Estimates of the cost of defensive expenditure to deal with long-term environmental damage have been made for the United Kingdom and Sweden amongst a number of other countries. They represent a significant cost component and are indicative of the pollution that results from the level of economic activity.[4,5] They are shown in Figure 3.

Air pollution costs are estimated on the basis of emission data for sulfur dioxide (SO_2), nitrogen oxides (NO_x), particulates, volatile organic compounds (VOCs) and carbon monoxide (CO). For the United Kingdom defensive expenditure on air pollution represents about 80 per cent of total defensive expenditures; in Sweden the percentage for air pollution does not reach 50 per cent of the total in any year and may be, in some years, slightly less than 30 per cent. This is due, in part, to large differences in national fuel mixes between the two countries.

Spectrum of Air Pollutants

The major air pollutants in industrialized or high-income countries arise mainly from combustion processes but also from industrial processes (chemical, metallurgical and construction), agricultural practices and abrasion and wear from vehicles. They include:

- carbon monoxide
- sulfur dioxide
- nitrogen oxides
- ammonia
- ozone
- fine particulates
- Volatile Organic Compounds (VOCs) including organic pollutants such as benzene and buta-1,3-diene
- heavy metals

These are by no means all of the significant pollutants emitted to air; exhaust gases from petrol engines may contain over 170 compounds not originally present in the petrol!

Air Pollution in Developing Countries

Developed countries use five times more energy on a per capita basis than less-developed countries. High-income countries of the World, with only 15 per cent of the World's population, use half its annual commercial energy.[7] This means that atmospheric pollution from the major sources of economic activity is far less in developing countries than in developed countries but three qualifications need to be borne in mind:

- large urban centres in developing countries have formidable air pollution

[7] IBRD, *World Development Indicators*, World Bank, Washington, DC, 2001.

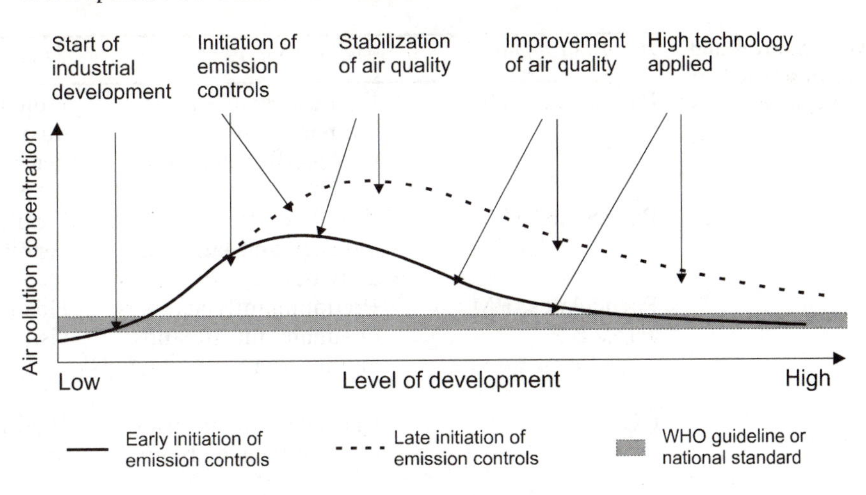

Figure 4 Development of air pollution problems in cities according to development status[8]

problems and do not match the defensive expenditure of the highly-developed countries – and by the year 2015 it is estimated that 8 of the 10 largest metropolitan areas in the World will be located in the less-developed World;

- in less-developed countries much of the primary energy consumed does not enter the commercial fuel market and is therefore difficult to estimate;
- the average annual total commercial energy use (1980–98) grew by only 1.7 per cent in high-income countries but by 4.4 and 5.1 per cent, respectively, in middle-income and low-income countries.[7]

One result of the rapid rate of growth in energy use is that some areas in the developing world received, for example, wet and dry deposition of sulfur compounds (mainly from emissions to the atmosphere from fossil fuel combustion) in 1985 that look very different from those predicted for 2050 (see Section 3, Figures 15 and 16). Predictions such as these emphasize the necessity for megacities in the developing World to speed-up the implementation of air pollution control measures and move from the stabilization of air quality to a strong improvement phase (Figure 4).

2 Impacts of Air Pollution

The progression of air pollution in different parts of the World has led to policy initiatives and defensive expenditures as a response to the impacts on the human and natural environment. Initial policy interventions in Europe were at urban scale to combat human health impacts (for example, the UK Clean Air Acts of 1956 and 1968) and later efforts have been directed towards reducing the emissions leading to acidification and eutrophication by long-range transboundary air pollution, causing impacts far from the sources of pollution. The impacts of pollution typical of Europe and North America are now being increasingly observed in low and middle income countries at both local and regional scales. However, the nature and spectrum of the problems differ between the different

[8] D. Mage, G. Ozolins, P. Peterson, A. Webster, R. Orthofer, V. Vandeweerd and M. Gwynne, Urban air pollution in megacities of the World. *Atmos. Environ.*, 1996, **30**, 681.

Table 1 Sources and effects of selected air pollutants discussed in this article

Polluting agent	*Major sources*	*Effects*
Particulates (TSP)	Fuel use: coal, oil, peat, biomass; construction process dust emissions	Human health; nuisance dust; soiling; ecosystem degradation; reduced crop quality and yield
Particulates (PM_{10})	Diesel vehicles; aerosols; industry and land use activities	Health effects; impact on visibility
Particulates ($PM_{2.5}$) aerosols	Predominantly aerosols of sulfate, nitrate and ammonium	Health effects; impact on visibility
CO	Energy use (combustion) of fossil fuels or biomass	Health effects; climate modification
NO_x (including nitrates from oxidation in the atmosphere)	Energy use (any high temperature combustion causes N to be fixed from the air) – any fuel, particularly in transport	Health effects; ecosystem acidification and eutrophication; precursor of photochemical oxidants (*e.g.* O_3)
SO_2 (and sulfate from oxidation in the atmosphere)	Energy use (combustion) of coal, oil; industrial processes (*e.g.* smelting)	Ecosystem acidification; health effects; impacts on crops, forests and natural vegetation; corrosion
NH_3/NH_4^+	Animal husbandry; fertilizer production and volatilization	Gaseous impacts on vegetation (by NH_3); acidification (by NH_4^+) and eutrophication
Volatile organic compounds	Road transport; solvent use; extraction and distillation of fossil fuels; non-combustion processes	Health impacts; precursor for photochemical oxidant pollution
Buta-1,3-diene	Vehicle exhausts	Health (carcinogen)
Benzene	Vehicle exhausts	Health (carcinogen)
Heavy metals	Energy use (combustion) of coal; metallurgical industry emissions	Human health effects; food chain effects; impacts on vegetation
O_3	Photochemical reaction between NO_x and VOCs in the presence of sunlight	Impacts on human health; impacts on crop yield, tree vitality and natural vegetation; corrosion

Table 2 Excess deaths associated with air pollution incidents (modified from ref. 9)

Date	*Place*	*Excess deaths*
December 1873	London	270–700
December 1892	London	1000
December 1930	Meuse Valley, Belgium	63
December 1952	London	4000
November 1953	New York, USA	250
December 1962	London	340–700
December 1962	Osaka, Japan	60

regions of the World. Table 1 shows the major air pollutants, their sources and impacts.

Local Air Pollution

Local air pollution is pollution in close proximity to urban centres and industrial installations, usually at a distance of less than fifty kilometres. Close to air pollution sources, in and around urban locations and close to industrial installations, there are impacts on human health, crop yield and quality, forest health, man-made materials and monuments, and also on visibility.

Human health. Human health impacts are caused by a wide range of gases and particles (Table 1). Sulfur dioxide, in combination with particulate matter, has caused serious health impacts in Europe and North America over the last 150 years (as indicated from the data in Table 2[9]). For example, there were a number of very bad smog episodes during the 1950s that led to the development of the Clean Air Act in the UK. Figure 5 shows the number of excess deaths related to the most infamous smog episode in London in 1952 where 4000 excess deaths were eventually recorded, which led to a public outcry forcing politicians to act.

Recent evidence suggests that fine particulates, less than 10 μm (PM_{10}) and, in particular, less than 2.5 μm ($PM_{2.5}$), size fractions that represent an inhalation hazard, are contributing to respiratory and cardiopulmonary disease[10] resulting in mortality and hospital admissions at pollutant concentrations well below air quality standards.[11] Indeed, epidemiological evidence from the United States is consistent with a linear non- threshold response for the population as a whole.[10] The PM_{10} and $PM_{2.5}$ concentrations are higher in parts of Asia than in Europe or North America and similar responses may be seen in these areas, although further epidemiological studies are required in developing countries to determine whether people in different socio-economic circumstances respond in the same or different ways to air pollutants.[12] However, as there is no evidence that a

[9] Modified from D. M. Elsom, *Atmospheric Pollution: A Global Problem*, Second Edition, Blackwell, Oxford, 1992.

[10] M. Lippman, The 1997 US EPA standards for particulate matter and ozone, in *Air Pollution and Health*, ed. R. E. Hester and R. M. Harrison, Issues in Environmental Science and Technology, No. 10, The Royal Society of Chemistry, Cambridge, 1998, pp. 75–99.

[11] F. Murray, Impacts on Health, in *Regional Air Pollution in Developing Countries*, Background Document for Policy Dialogue. Stockholm Environment Institute, York, 1998, pp. 5–11.

[12] F. Murray, G. McGranahan and J. C. I. Kuylenstierna. The application of models to assess health effects of air pollution in Asia, *Water, Air, Soil Pollut.* (in press).

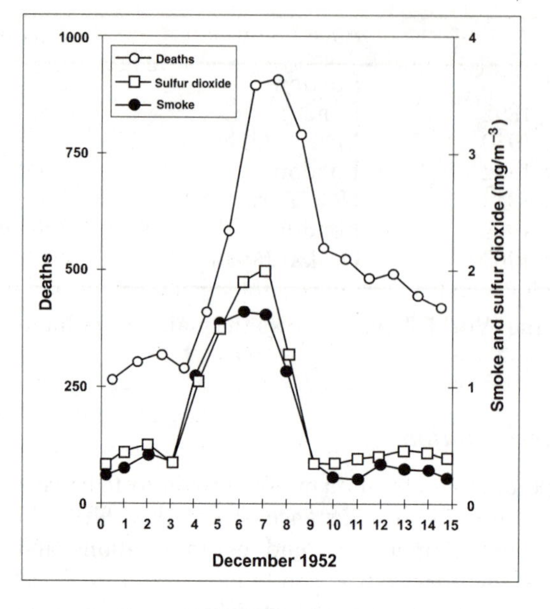

Figure 5 Deaths related to sulfur dioxide and smoke concentration in the London smog, December 1952[13]

threshold to PM-associated health effects exists it may be reasonable to expect that reductions in PM_{10} and $PM_{2.5}$ will result in improvements in peoples health.[14]

Oxides of nitrogen are associated with increased incidence of lower respiratory tract infection in children and decreased airway responsiveness in asthmatics.[15] It has proved difficult to disentangle the impact of outdoor NO_x impacts from indoor, and from the presence of other pollutants, but published estimates suggest respiratory effects in children at annual average NO_2 concentrations in the range 50–75 $\mu g\, m^{-3}$ or higher.[16]

Ozone-related health effects include changes in lung capacity, flow resistance, epithelial permeability and reactivity to bronchoactive challenges. Short term effects on lung function can be observed a few hours after the exposure and may persist for hours or days after the exposure ceases.[10] These effects are apparently reversible, but repeated exposure can exacerbate or prolong these effects, although the current knowledge of the chronic health effects are much less complete.[10] Field studies also show decreased athletic performance and increased incidence of asthma attacks and respiratory symptoms in asthmatics.[11]

There has been a steady improvement in air quality in cities of Europe since the 1970s due to policies that reduce pollutants such as sulfur dioxide and particulates, and excess death rates have fallen. In Asia, some urban areas seem to be following the same pollution trajectory as cities such as London up until the

[13] E. T. Wilkins, Air pollution and aspects of the London fog of December 1952, *Quart J. R. Met. Soc.*, 1954, **80**, 267.

[14] M. Lippman, Air Pollution and Health – Studies in North America and Europe, in *Health and Air Pollution in Developing Countries*, ed. G. McGranahan and F. Murray, Stockholm Environment Institute, York, 1999, pp. 29–41.

[15] L. J. Folinsbee, Human health effects of air pollution, *Environ. Health Perspect.*, 1992, **100**, 45–56.

[16] WHO, *Updating and Revision of the Air Quality Guidelines for Europe*, Reports of WHO Working Groups, WHO Regional Office for Europe, Copenhagen, 1995.

1950s. In China annual average SO_2 concentrations of 402 $\mu g\,m^{-3}$ and particulate matter (TSP) concentrations of 690 $\mu g\,m^{-3}$ have been recorded in 'urban' sites in Chongqing in 1985–89.[17] Globally more than 1200 million people may be exposed to excessive levels of sulfur dioxide and more than 1400 million people to excessive levels of suspended particulate matter.[18]

Specifically, some traffic-related pollution (giving rise to oxides of nitrogen, CO, particulates and VOCs) has continued to increase in many parts of the world as vehicle numbers increase, often negating improvements in engine or fuel technology. Lead in petrol has been phased out in most of Europe, and the health damaging lead concentrations have been reduced. Focus in Europe has now been placed on reducing PM_{10}, NO_x, benzene and buta-1,3-diene (both carcinogens) and plans are being implemented to reduce these pollutants.[19] The nature of the pollution in urban areas in Europe is therefore changing, even though the overall impact has reduced. Air pollution from the transport sector now dominates in some developing country cities, where particulate emissions from diesel vehicles can be very high, leaded petrol is still in use and where hot, sunny conditions readily give rise to photochemical smogs, leading to increased ozone- and NO_x-related health impacts.

Indoor air pollution is a severe problem in many developing countries, particularly in many parts of Africa and Asia. This is especially so for women, cooking over open fires in houses that have poor ventilation. It is a major problem in rural areas and villages, where the poor may be using dung or fuelwood for cooking, rather than cleaner fuels such as kerosene or gas. Roughly 20–35 per cent of total energy consumption in developing countries uses 'traditional fuels' of wood and other biomass fuels.[20] However, the scale of the health impact of indoor air pollution is difficult to estimate as there are relatively few data, but it has been estimated[21] that 2.8 million premature deaths per annum may be the result of indoor air pollution.

Biomass burning is an air pollution problem more clearly related to Asia, Africa and Latin America than with Europe or North America. The total emission of particulate matter and nitrogen oxides can be high, but generally spread over large areas. The smoke haze episode of 1997 in Indonesia and surrounding countries was related to fires used for forest clearance burning out of control. This became a regional health problem with air quality indices reaching hazardous levels and states of emergency declared in areas in several countries in South-East Asia. More than 8000 people were admitted to hospital in Malaysia with health problems related to air pollution.[11]

[17] T. Larssen, H.-M. Seip, A. Semb, J. Mulder, I. P. Muniz, R. D. Vogt, E. Lydersen, V. Angell, T. Dagang and O. Eilertsen, Acid deposition and its effects in China: an overview, *Environ. Sci. Policy*, 1999, **2**, 9–24.

[18] UNEP, *Urban Air Pollution*. UNEP/GEMS Environment Library No. 4, United Nations Environment Programme, Nairobi, Kenya, 1991.

[19] DETR, *Air Quality and Transport*. Part IV The Environment Act 1995 Local Air Quality Management, Department of the Environment, Transport and the Regions, London, 2000.

[20] WRI, *World Resources: 1998–99*, World Resources Institute. Oxford University Press, Oxford, 1998.

[21] K. R. Smith, Indoor air pollution in developing countries: growing evidence of its role in the global disease burden, in *Proceedings of Indoor Air '96*, 7th International Conference on Indoor Air Quality and Climate, Institute of Public Health, Tokyo, 1996.

The cost of air pollution to developing country cities can be very high. Using dose–response relationships developed for health and particulate matter, the URBAIR project calculated the costs of air pollution impacts on health in Mumbai, Metro Manilla, Jakarta and the Kathmandu Valley caused by PM_{10} exposure.[22] The cost of excess deaths was calculated using the human capital approach and chronic impacts based upon restricted activity days, asthma attacks and respiratory symptom days. The health costs in the these cities varied from US $2.84 million in Kathmandu to US $127 million per annum in Jakarta, with Mumbai and Manila also showing health costs greater than US $100 million per annum.

Air quality guidelines to protect human health have been produced by international organizations such as the World Health Organisation (WHO), regional organizations, such as EU and national governments (see Table 3 for some examples). When comparing pollutant concentrations in cities of Asia to air quality guidelines, the concentrations of particulate matter consistently register as being of serious concern. For example, concentrations in Bangkok, Jakarta, Manila, Beijing, Delhi and Seoul exceeded WHO guidelines by more than a factor of two in 1992.[23] In addition, SO_2 pollution in NE Asian cities, such as Beijing and Seoul, also indicates a serious situation. Ozone causes 'moderate to heavy' pollution in Jakarta and Beijing, showing similar exceedance of WHO guidelines (up to a factor of two) to New York or Tokyo.[23]

Impacts of gaseous air pollution on vegetation. Ozone and sulfur dioxide have impacts on crops, forests and natural vegetation in both industrialized and developing countries. The direct impacts of elevated SO_2 concentrations are largely confined to urban areas, peri-urban areas and close to industrial sources of pollution, although dry and wet deposition of resulting acidifying substances occurs on a continental scale (see Section 2). Ozone, once formed, may be transported in the atmosphere and affect much larger areas. Gaseous air pollutants may affect vegetation through visible injury and/or effects on growth and yield (invisible injury) and through subtle physiological, chemical or anatomical changes.[28] Visible damage to leaves may have direct impacts on the market value of crops such as spinach and tobacco. In the 1960s it was widely believed that yield could only be affected when visible injury was present, but a large body of knowledge has now shown that significant yield losses can occur in

[22] World Bank, *Urban Air Quality Management Strategy in Asia: Guidebook*, World Bank (ISBN: 0-8213-4032-8). Washington, DC, 1997.

[23] ASEAN, First ASEAN State of the Environment Report. ASEAN Secretariat, Jakarta, 1997.

[24] Council of Ministers of the European Commission, *Air Quality Assessment and Management*, Report 96/62/EC, Brussels, 1996.

[25] US EPA, *National Primary and Secondary Ambient Air Quality Standards*, Air Programs 40CFR Part 50, Office of Air Quality Planning and Standards, Research Triangle Park, NC, 1997.

[26] US EPA, *EAP's Clean Air Standards – A Common Sense Primer*, Office of Air Quality Planning and Standards, Research Triangle Park, NC, 1997.

[27] H. Fujimaki, Key air pollutants in Japan, in *Global Air Quality Guidelines*, World Health Organisation, Geneva, 1998.

[28] M. R. Ashmore, Air pollution and agriculture, *Outlook on Agriculture*, 1991, **20** (3), 139–144.

Table 3 International guideline values for some urban air pollutants as a time-weighted average concentration in air[12]

	European Union air quality standards[a]		*US air quality standards*		*Japanese air quality standards*		*WHO air quality Guidelines for Europe*[d]	
Pollutant	*Concentration* $\mu g\,m^{-3}$	*Averaging time*	*Concentration* $\mu g\,m^{-3}$	*Averaging time*	*Concentration* $\mu g\,m^{-3}$	*Averaging time*	*Concentration* $\mu g\,m^{-3}$	*Averaging time*
SO_2	350[b]	1 hour, 99.7 percentile	267	1 hour	267	1 hour	500	10 mins
	125[b]	24 hour, 99 percentile	107	24 hour	107	24 hour	125	24 hour
	20	Annual					50	Annual
NO_2	200[c]	1 hour, 99.9 percentile					200	1 hour
	40[c]	Annual	100	Annual	76–115	24 hour	40–50	Annual
O_3	160	1 hour	240	1 hour	120	1 hour		
	120	8 hour	160	8 hour[e]			120	8 hours
Particulate								[f]
TSP	200	1 hour			200	1 hour		
	100	24 hour			100	24 hour		
PM_{10}	50[b]	24 hour, 98 percentile	150	24 hour				
	100[b]	Annual	50	Annual				
$PM_{2.5}$			50	24 hour				
			15	Annual				

[a] Proposed by the Commission.[24]
[b] Takes effect from 1 January, 2005.[25,26] [c] Takes effect from 1 January, 2010.[27]
[d] The guideline values should be read in the context of the guideline documents.[16]
[e] Proposed goal.
[f] No guideline values were set for particulate matter because there is no evident threshold for effects on morbidity and mortality.

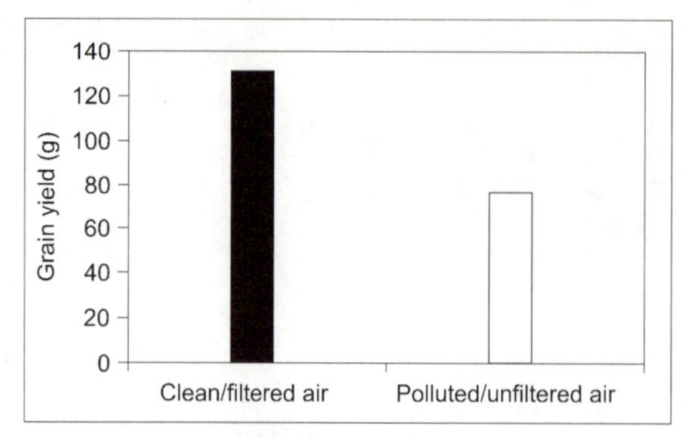

Figure 6 Total filled grain yield (dry weight per plant in g) of Basmati rice (cv. Basmati-385) at final harvest for plants grown in open-top chambers in Pakistan (7 km south of Lahore) in air filtered of pollutants compared to unfiltered (polluted) air showing a 42% reduction in grain yield[29]

the absence of visible symptoms. Most work on yield reductions has taken place in Europe and North America, but work in Asia is increasingly showing significant yield reductions in crops such as rice (Figure 6) and wheat at ambient pollutant concentrations in comparison to air filtered of pollutants. As concentrations of NO_x and VOCs increase in line with increasing vehicle use in developing countries, tropospheric ozone concentrations are liable to increase and the potential for yield losses will become greater.

Crop loss assessments to date have concentrated on the direct impacts of air pollution on yield, and have not taken into account effects on crop quality or the indirect impacts on yield. Reductions in income for vegetable producers and suppliers can arise from visible damage to the edible portion of the crop. In addition, there are other potential non-visible impacts of air pollution such as reductions in nutritional quality or accumulation of heavy metals, with important implications for consumers, particularly the poor.[30,31]

Corrosion impacts on materials. Corrosion of materials has mainly been a topic studied in Europe and North America. However, the pollution levels in Asia have increased rapidly and, as many developing countries happen to be located in warm, humid regions with high relative humidity and high frequency of rainfall, there is a great risk of extreme corrosion rates, even higher than in temperate zones at the same pollutant concentration.[32] From a comparison of corrosion data from China and Europe, the sensitivity to SO_2 is similar in tropical climates to that in temperate, but the sensitivity to acidic wet deposition is much higher in wet tropical conditions. For non-marine sites Chinese data (Figure 7) show that corrosion rates are 4–5 times higher for carbon steel and 2–3 times higher for zinc and copper in Chinese sites than in UN/ECE test sites in Europe. This is due

[29] A. Wahid, R. Maggs, S. R. A. Shamsi, J. N. B. Bell and M. R. Ashmore, Effects of air pollution in rice yield in the Pakistan Punjab. *Environ. Pollut.*, 1995, **90**, 323.

[30] F. Marshall, M. Ashmore and F. Hinchcliffe, *A Hidden Threat to Food Production: Air Pollution and Agriculture in the Developing World*, International Institute For Environment and Development, London, 1997.

[31] M. R. Ashmore and F. M. Marshall, Ozone Impacts on Agriculture: An Issue of Global Concern. *Adv. Bot. Res.*, 1999, **29**, 32–52.

[32] J. Tidblad, A. A. Mikhailov, and V. Kucera, *Acid Deposition Effects on Materials in Subtropical and Tropical Climates.* Data compilation and temperate climate comparison, Swedish Corrosion Institute KI Report 2000:8E, Stockholm.

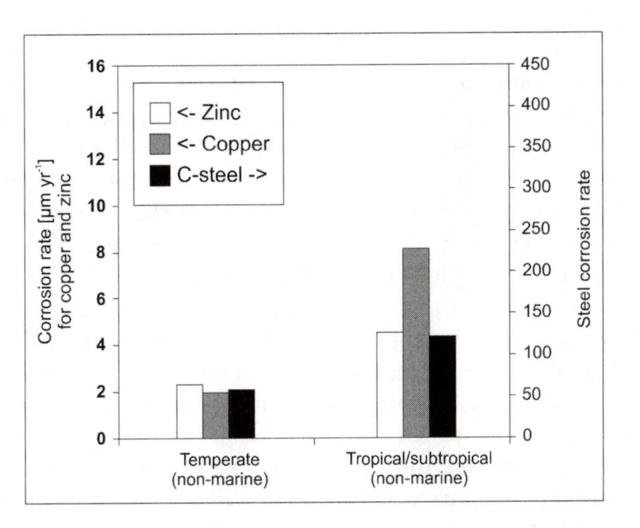

Figure 7 Maximum of observed corrosion rates for zinc, copper and C-steel for temperate and tropical/sub-tropical climates[32]

partly to the higher SO_2 concentrations that now exist in China and partly due to climatic differences increasing the sensitivity of materials in tropical and sub-tropical conditions. However, the influence of the climate on dose–response relationships is not fully known and the data for those dose–response functions that exist (for China) are limited and require verification. The Chinese dose–response relationships indicate that although the response to dry deposition is similar to the European data, the corrosion rates for wet deposition are much higher for copper and zinc using the Chinese relationships. Several climatic differences in wet sub-tropical and tropical regions, compared to temperate climates, need to be taken into account in order to quantify corrosion effects:[32]

(i) the higher temperature and solar radiation;
(ii) high relative humidity throughout the year;
(iii) the potential for increased dew formation (due to large differences between day- and night-time temperatures);
(iv) the different character of rain in these regions.

Impacts on visibility. There exists a close association between the concentration of particles in the atmosphere, their light scattering coefficient (properties) and the range of visibility. Although more complex relationships exist,[33] a robust relationship can be defined:[34]

$$V = 884.8/M$$
V = visibility in miles at noon in dry conditions
M = mass of particulate matter ($\mu g\, m^{-3}$)

[33] G. Landrieu, *Visibility Impairment by Secondary Ammonium, Sulphates, Nitrates, and Organic Particles*, Draft Note Prepared for the UN/ECE Convention on LRTAP, Copenhagen 9–10 June, 1997.

[34] K. Noll *et al.*, Visibility and aerosol concentration in urban air, *Atmos. Environ.*, 1968, **2**, 465–475.

This stresses the direct relationship between particulate matter concentrations and visibility. According to Maddison,[35] visibility range in Europe is a 'good' that has significant economic value, although little attention to such valuation has occurred in Europe. Visibility can never really be separated from the aesthetic qualities of the landscape, which casts doubt on the transferability of valuation exercises. However, many developing countries rely on visibility for tourism. In Nepal, atmospheric visibility data indicate that there has been a substantial decrease in visibility in the Kathmandu Valley since 1970. The number of days with good visibility in Winter at Kathmandu Airport has decreased from 25 days per month in 1972 to 5 days per month in 1992.[36] The bowl-like topography and low wind speeds during the winter season create poor dispersion conditions, predisposing the Kathmandu Valley to serious air pollution problems[22] (World Bank). A pessimistic, but accurate image of the air pollution situation in the Kathmandu Valley[37] gave negative publicity to the area that could have had an adverse impact on tourism. In the early 1990s, foreign currency revenues amounted to approximately US$60 million a year. Although no 'dose–effect' relationships of air pollution and tourism are available, if it were assumed that there could be approximately 10% decrease in tourism, then this could lead to a loss of to US$6 million for Nepal. This is a very significant amount of foreign exchange for a country that has a negative balance of trade.[22]

Regional Air Pollution

Gaseous emissions such as sulfur dioxide and nitrogen oxides are rapidly oxidized to sulfate and nitrate, and ammonia is transformed to ammonium in the atmosphere. These pollutants can travel over very long distances, transported by the winds over hundreds of kilometres and then deposited by wet deposition in rain, occult (in cloud) deposition and dry deposition to surfaces and may cause impacts far from the source of pollution. Sulfur and nitrogen deposition may acidify ecosystems and nitrogen may be responsible for eutrophication (over-fertilization) of terrestrial ecosystems. Ozone is another gas which, once produced, may be transported over long distances and have effects on vegetation. Acidification, to a lesser extent eutrophication, and more recently ozone impacts on crops and forests are impacts that have led to the development of negotiations between countries in Europe and elsewhere that are receiving each others' pollution. The need for such negotiations becomes clear when one considers that more than 90% of the acidic deposition in Norway (where acidification of lakes has been a large problem) came from sources outside Norway during the early 1980s (EMEP model calculations[38]) when the negotiations were gaining momentum.

[35] D. Maddison, *The Economic Value of Visibility: A Survey*, Centre for Social and Economic Research on the Global Environment (CSERGE), University College, London and University of East Anglia, 1997.

[36] UNEP, *State of the Environment: Nepal*, United Nations Environment Programme, Regional Resource Centre for the Asia-Pacific. Pathumthani, Thailand, 2001.

[37] Article in *Newsweek*, October 1993.

[38] J. Lehmhaus, J. Saltbones and A. Eliassen, *Deposition Patterns and Transport Sector Analyses for a Four-Year Period*, EMEP/MSC-W Report 1/85, The Norwegian Meteorological Institute, Oslo, 1985.

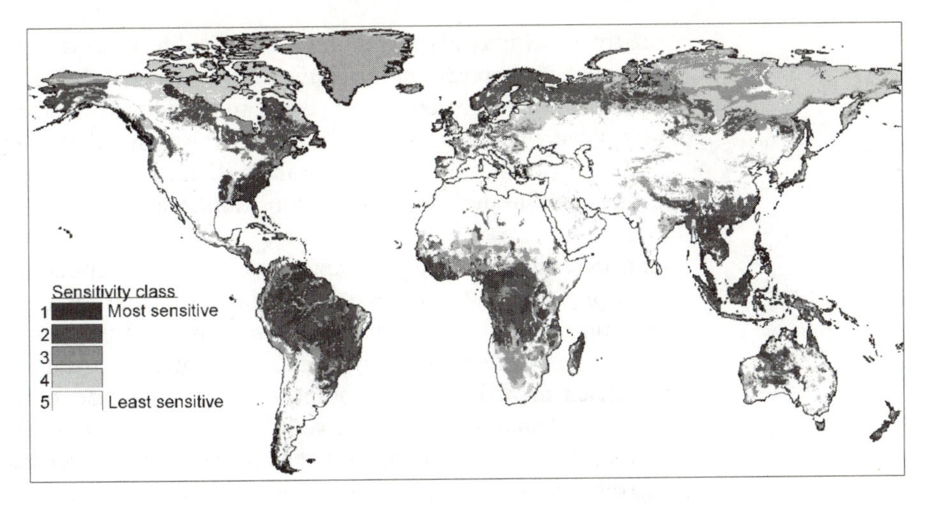

Figure 8 The distribution of relative ecosystem sensitivity to acidic deposition based upon the soil buffering characteristics[41]

Acidification. Soil acidification occurs when base cations on the soil exchange complex are replaced by hydrogen and/or aluminium. Base cations are supplied by weathering and atmospheric deposition and removed by plant uptake and leaching. Leaching of cations depends on leaching of mobile anions such as SO_4^{2-} and NO_3^-. In many areas input and output of SO_4^{2-} are quite similar. However, adsorption and/or plant uptake may occur. The nitrogen processes are more complex. Atmospheric deposition includes both NO_x and NH_4^+; the latter may give NO_3^- and hydrogen ions through nitrification. Nitrogen-saturated ecosystems (where nitrogen inputs are in excess of ecosystem nutrient needs,[39] such as seen in highly polluted parts of Europe) will leach a greater proportion of nitrogen deposited than ecosystems where nitrogen is limiting vegetation growth. In some areas, very little nitrate leaches at all. As bases leach from the soil, the base saturation and pH may decrease (the soil acidifies) if the soil is not sufficiently buffered by soil mineral weathering. The rate of acidification will depend on the capacity of the base cation storage on the cation exchange complex. Lake and stream water acidification occurs as the soils in the catchment acidify and the decrease in pH, increases in aluminium concentrations and the loss of fish have been the clearest and most severe impacts of acidic deposition, particularly in the sensitive regions of Scandinavia, the UK and NE North America. The soil acidification can itself have impacts on the vegetation at a site, leading to losses in biodiversity and plant vigour. The widespread tree damage in central Europe and parts of Canada and the USA have been attributed to 'acid rain'. The indications are that pollution is one of the main causes, but that the situation relating to tree damage is the result of a complex interaction of different pollutants with biotic and climatic stresses.[40]

Kuylenstierna *et al.*[41] have used soil buffering characteristics to map the

[39] J. D. Aber, W. McDowell, K. Nadelhoffer, A. Magill, G. Berntson, M. Kamakea, S. McNulty, W. Currie, L. Rustad and I. Fernandez, Nitrogen saturation in temperate forest ecosystems, hypotheses revisited. *BioScience*, 1998, **48**, 921–934.

[40] A. Wellburn, *Air Pollution and Acid Rain: The Biological Impact*, Longman, Harlow, 1990.

[41] J. C. I. Kuylenstierna, H. Rodhe, S. Cinderby and K. Hicks, Acidification in developing countries:

relative sensitivity of different regions to acidification-related impacts (Figure 8). Although concentration on acidification in Asia has only recently begun, several studies have indicated evidence for soil acidification in China that may be associated with acidic deposition.[42] Figure 8 shows that many ecosystems in NE and SE Asia may be very sensitive and so the risks of acidification damage may be rather high in areas with high deposition rates.

Eutrophication. Nitrogen deposition, either as ammonium or nitrate, can eutrophy (nutrify to excess) terrestrial ecosystems, causing changes to both structure and function. Nitrogen is an important plant nutrient, often limiting growth in terrestrial ecosystems. Therefore, nitrogen additions can cause increased growth which promotes the proliferation of species with a high nitrogen demand at the expense of those species requiring less nitrogen. As Tamm[43] notes, most of the threatened (rare) species in central Europe have a low nitrogen demand and therefore are most at risk from widespread increases in nitrogen deposition. In developing countries, nitrogen has not been as clearly studied, but many florally diverse ecosystems could be at risk if nitrogen deposition increases.

Although freshwater ecosystems are often phosphorus limited, coastal marine ecosystems can be very sensitive to increasing concentrations of nitrate. This applies also in tropical regions where coral reefs may be significantly affected. The majority of nitrate in rivers comes from the agricultural sector, but a significant proportion comes from atmospheric deposition in some areas, such as in the Baltic Sea where 20% of inputs derived from the atmosphere.[44]

3 The Regional Status of Air Pollution

Driving Forces

Many developing countries have experienced progressive degradation in air quality over the past decades. Rapid urbanization, increased industrialization and rising energy use, mostly derived from fossil fuel combustion, have been the major contributors. For example, energy demand in Asia and the Pacific, which has 30 per cent of the World's land area supporting 60% of the World population, is growing faster than in any other part of the World.[45]

Since 1950, the number of people living in urban areas has risen from 750 million to more than 2500 million people.[45] The Asian and Pacific region has experienced the greatest absolute increase in urban population, with the urban population almost doubling between 1975 and 1995 and the trend for the growth

ecosystem sensitivity and the critical load approach on a global scale. *Ambio*, 2001, **30**, 20–28.

[42] Dai Zhaohua, Liu Yunxia, Wang Xinjun and Zhao Dianwu, Changes in pH, CEC and exchangeable acidity of some forest soils in southern China during the last 32–35 years, *Water, Air, Soil Pollut.*, 1998, **108**, 377–390.

[43] C. O. Tamm, Nitrogen in Terrestrial Systems. *Ecological Studies*, **81**. Springer-Verlag, 1991.

[44] M. Enell and J. Fejes, The nitrogen load to the Baltic Sea – present situation, acceptable future load and suggestion source reduction, *Water, Air, Soil Pollut.*, 1995, **85**, 877–882.

[45] UNEP, *Global Environment Outlook – 2.* United Nations Environmental Programme, Global State of the Environment Report 1997, Oxford University Press, New York, 2000.

of megacities in the region increases the likelihood of environmental and social stresses. Rapid increases in urbanization are also occurring in Latin America and Africa, albeit rather more slowly compared with Asia. However, it is worth noting that the percentage of the population in urban areas is still highest in North America and Europe.

Increasing populations inevitably mean increased use of motor vehicles. Unlike European and American cities in the 1950s, motor vehicles now account for the major share of pollutant emissions in cities such as Delhi (57%), Beijing (75%), Manila (70%) and Kuala Lumpur (86%).[46] The Asia and Pacific region experienced an increase in vehicle numbers of 40% between 1980 and 1995. This represents a greater rate of increase than in North America, though not as large as increases experienced in Europe. If current rates of expansion continue there will be more than 1000 million vehicles on the road world-wide by 2025.[45]

Trends in Sources and Emissions of Pollution

Air pollution from industry has increased in a number of developing countries. Although the sizes of industrial plants have tended to be relatively small by industrialized country standards, the cumulative effect of many small industrial sources of pollution is considerable. In addition, the displacement of polluting industries to countries where less emphasis is placed on emission control could cause significant regional problems in the future.[47]

The growth in energy demand in recent decades has been particularly marked in rapidly industrializing countries and regions. Between 1990 and 1993 Asia's energy consumption grew by 6.2% per annum, whilst the global energy consumption fell by 1%.[48] The Asia-Pacific region accounted for 41% of world coal consumption in 1993[49] and large power plants are the greatest contributors to sulfur dioxide pollution in many developing country regions. Figure 9 shows that at the same time as SO_2 emissions have been decreasing in Europe and North America, the increased combustion of fossil fuels in Asia has resulted in increased SO_2 emissions; these trends are projected to continue over the next decades.

If developing countries follow a conventional development path, with the heavy reliance on coal and oil, with convergence of the use of control measures to OECD 1995 levels by 2025, large increases in emissions of SO_2 and NO_x will result in some regions. As shown in Figures 10 and 11 the largest increases would be seen in China and South and South-East Asia.

Increasing populations and changing eating patterns necessitate expansion in the agricultural sector with increases in animal numbers and fertilizer use. Figure

[46] World Resources Institute, *World Resources 1994–1995*, A Joint Report by WRI, UNEP and UNDP, Oxford University Press, New York, 1994.

[47] N. Abdul Rahim, General problems associated with air pollution in developing countries, in *Air Pollution and the Forests of Developing and Rapidly Industrializing Countries*, eds. J. L. Innes and A. H. Haron, IUFRO Research Series 4, CABI Publishing, Oxon., UK, 2000.

[48] ADB, *Emerging Asia: Changes and Challenges*. Asian Development Bank, Manila, The Philippines, 1997.

[49] EIA, *International Energy Annual: 1993*, Energy Information Agency, US Department of Energy, Washington, DC, 1995.

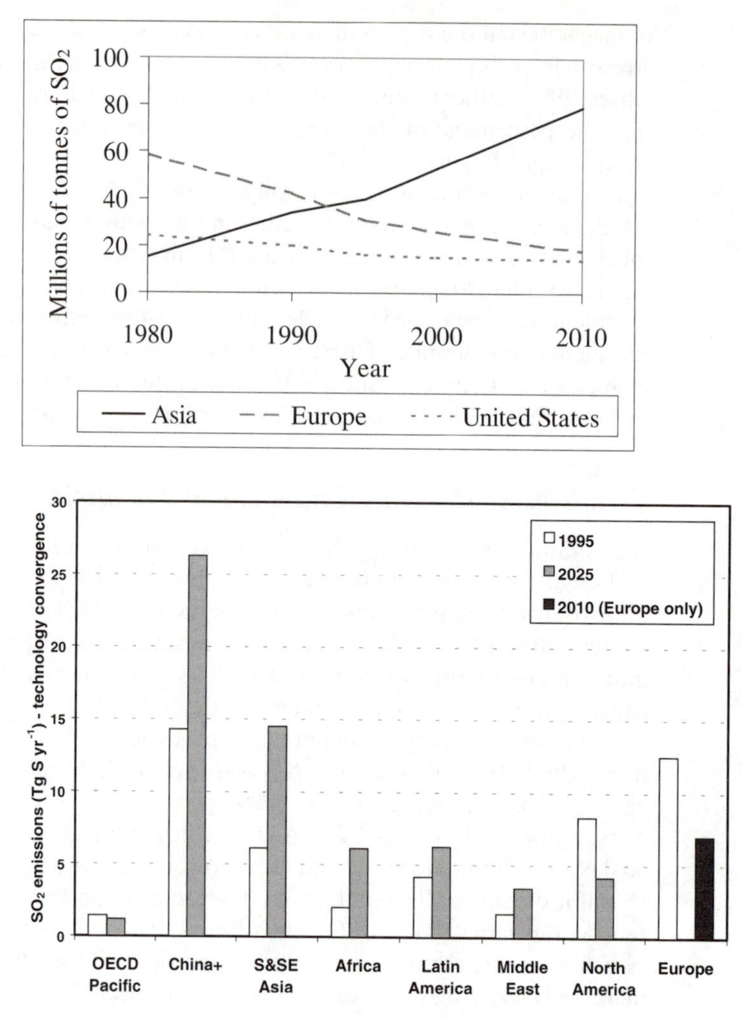

Figure 9 Regional SO_2 emissions from fossil fuel burning (from data in ref. 45)

Figure 10 Anthropogenic SO_2 emissions for 1995 and projections for 2025 assuming a conventional development with convergence of control technology by 2025 in developing countries to OECD 1995 levels[50]

12 shows the increases in ammonia emissions that could result if the management of such emissions is not improved.

Air pollution trends in Asia and the Pacific. The per capita commercial energy use more than doubled in most parts of the Asia-Pacific region between 1975 and 1995.[45] Fossil fuels account for approximately 80% of energy generation in the region with both China and India relying heavily on coal,[49] and this has resulted in rapidly increasing emissions of sulfur dioxide (at a rate at least four times higher than any other region between 1970 and 1986[51]). NO_x emissions from fossil fuel combustion have also increased (by about 70%).[51]

[50] H. W. Vallack, S. Cinderby, J. C. I. Kuylenstierna and C. Heaps, Emission inventories for SO_2 and NO_x in developing country regions in 1995 with projected emissions for 2025. *Water, Air, Soil Pollut.*, (in press).

[51] S. Hameed and J. Dignon, Global emissions of nitrogen and sulfur oxides in fossil fuel combustion 1970–86, *J. Air Waste Manage. Assoc.*, 1992, **42**, 159–163.

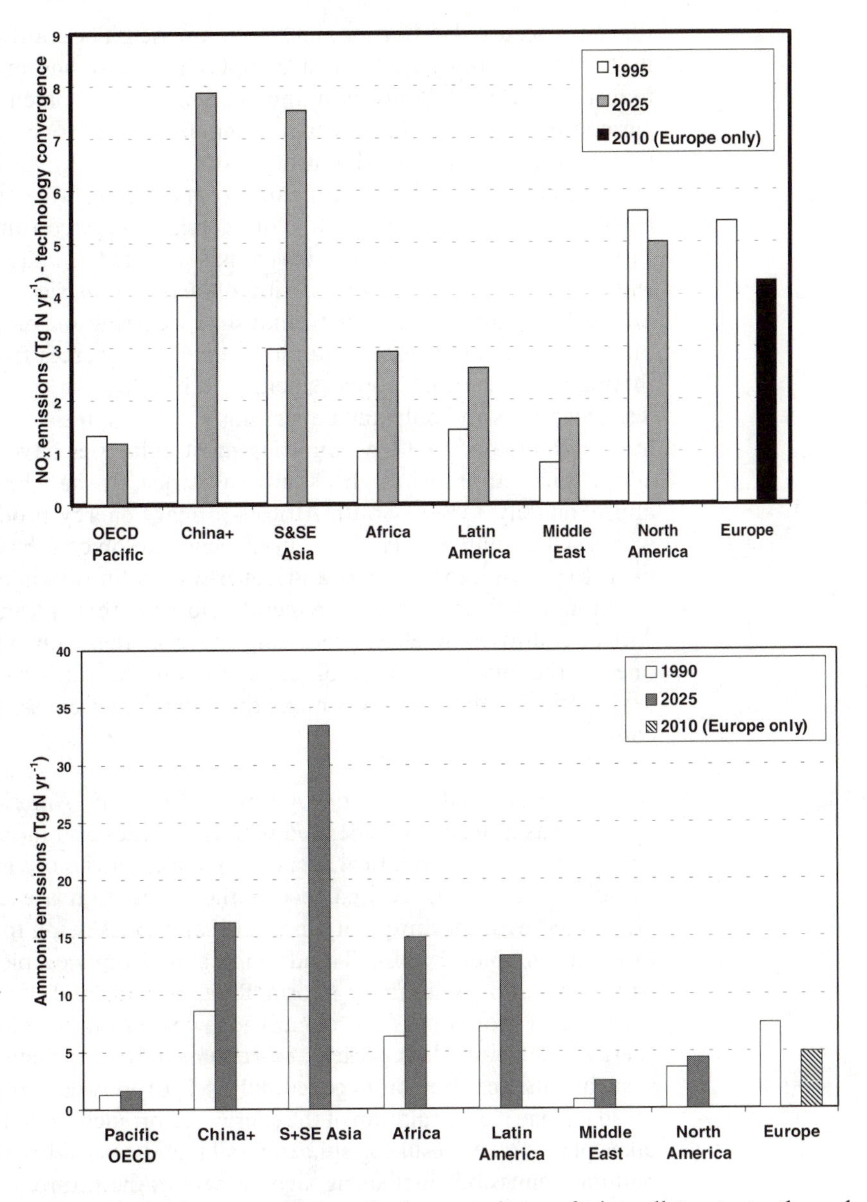

Figure 11 Anthropogenic NO_x emissions for 1995 and projections for 2025 assuming a conventional development pathway with convergence of control technology by 2025 in developing countries to OECD 1995 levels[50]

Figure 12 The EDGAR NH_3 emissions estimates for 1990 (anthropogenic + natural) and projected estimates for 2025 assuming a conventional development pathway and no further improvements in emission control techniques for anthropogenic sources[52]

Transportation contributes the largest share of air pollutants to the urban environment; urban levels of smoke and dust are generally twice the world average and more than five times as high as in industrial countries and Latin America.[48]

Air pollution trends in Africa. Industrial emissions of air pollutants in Africa only started in the mid-20th century and impacts are not yet as serious as those

[52] H. W. Vallack, S. Cinderby, C. Heaps, J. C. I. Kuylenstierna and K. Hicks, *Emissions Inventories for Anthropogenic SO_2 and NO_x* in Developing Country Regions in *1995 and Projected Emissions of SO_2, NO_x and NH_3 for 2025 and 2050*, Stockholm Environment Institute, York, 2001.

that have occurred in Europe and North America. The continent has transformed itself from a rural society to a complex one that has made great strides in industrialization, urbanization and economic development. Industries such as agriculture, mining, forestry and manufacturing, together with a growing population, have combined to bring about some environmental problems which were virtually non-existent at the turn of the 20th century. Since 1973 there has been a 145% increase in Africa's commercial energy consumption from 89.7 to 219.2 Mt of oil equivalent[53] which has resulted in large increases in SO_2 emissions. The highest concentration of heavy industries are found in Zambia, South Africa and Nigeria, with industry occurring on much smaller scales in other African countries.[54] The major emission sources from industry include thermal power stations, copper smelters, ferro-alloy works, steel works, foundries, fertilizer plants and pulp and paper mills.[45] Although cleaner production centres have been created in a few countries, most industries have made little effort to adopt such approaches. In southern Africa, it has been reported[55] that approximately 72% of South Africa's primary energy production is from coal with a sulfur content of *ca.* 1%. This dependence on coal-based thermal power is likely to persist into the future and hence SO_2 pollution will remain a problem. In addition, the World's richest mineral field runs through most of the southern African countries. Smelters processing ores from these mineral deposits represent one of the major sources of air pollution in southern Africa and future exploitation of these deposits make this industry an emission sector of growing concern.

Air pollution trends in Latin America. In Latin America air pollution is perceived as a problem associated with large cities and industrial areas but it is not yet prominent in political agendas. In the countries of Latin America and the Caribbean, nearly three quarters of the population are urbanized, many in megacities, with the air quality in most major cities rising to levels which pose a threat to human health. Trends emerging from completed (Uruguay and Argentina) and preliminary (Costa Rica, Mexico and Venezuela) inventories suggest that more than 50% of emissions come from industrial production and energy generation.[44] At present, more than 50% of the energy produced in the Central American region is generated by hydropower. However, this situation could change if deregulation of the energy sector, such as that suggested in Brazil, takes place.[56] The resulting shift to fossil fuel use would result in energy-related pollutant emissions increasing significantly in the future.

In Latin America there is another important source of air pollution originating

[53] J. McCormick, *Acid Earth – The Politics of Acid Pollution*, 3rd Edition, Earthscan Publications, London, 1997.

[54] A. M. van Tienhoven, K. A. Olrich and T. F. Fameso, Forestry Problems in Africa, in *Air Pollution in and the Forests of Developing and Rapidly Industrialising Countries*, ed. J. L. Innes and A. H. Haron, IUFRO, Cambridge University Press, Cambridge, 2000, pp. 79–99.

[55] G. Held, B. J. Gore, A. D. Surridge, G. R. Tosen, C. R. Turner and R. D. Walmsley (eds.), *Air Pollution and its Impacts on the South African Highveld*, Environmental Scientific Association, Cleveland, 144 pp.

[56] L. P. Rosa and R. Schaeffer, Global Warming Potentials: the case of emissions from dams, *Energy Policy*, 1996, **23**(2), 149–158.

from biomass burning. Large amounts of air pollution can be produced by the use of fuelwood, widespread agricultural practices (*e.g.* burning of sugar cane residues), the common practice of deforestation by burning, burning to renew pastures and large-scale uncontrolled vegetation fires (*e.g.* the vegetation fire in Roraima, in the Brazilian Amazon, in 1998).

Air pollution trends in Europe and North America. In Europe and North America, the impacts of pollution by sulfur dioxide (SO_2) and nitrogen oxides (NO_x) have led to agreements to reduce emissions of these gases. Consequently, SO_2 emissions are declining in these regions and in 1995 were 48% lower in Western Europe and 32% lower in North America compared with 1980 levels.[57] Emissions of NO_x have stabilized and are reducing slightly in these regions with 1995 levels 16% lower in Europe and 6% lower in North America compared with 1989 levels (when NO_x emissions peaked).[57]

Assessing Air Pollution Impacts

Information on trends in emissions is useful, but limited, as it does not give a comprehensive view of the potential changes in the impacts caused by air pollution. For this reason, various methodologies have been developed to allow assessments of the impacts associated with emission changes. The essential ingredients are an atmospheric transfer model for the appropriate scale which can use the emission estimates and model pollutant concentrations and depositions and then, through an understanding of causal relationships, link levels of pollution to impacts. There are essentially two ways in which this is carried out: to use dose–response relationships or to use threshold values. Concentration or deposition values can be used in combination with dose–response relationships to estimate the magnitude of the response or can be used to show when and where they exceed threshold values.

For health impacts at urban scale recent studies relating to the occurrence of daily deaths (total and by cause) to daily changes in air pollution levels have provided strong evidence of the health effects associated with particulate pollution. A pooled estimate of major studies suggested that a 10 mg m^{-3} increase in PM_{10} will be associated with an increase in daily mortality equal to 0.74%.[58] Similarly, an increase of 10 mg m^{-3} in $PM_{2.5}$ levels will increase mortality by 1.5%.[58] Using such information a recent estimate for Delhi suggests that an annual reduction of 100 mg m^{-3} in TSP could be associated with a reduction of about 1400 premature deaths per year.[59] Similar response information for ozone exists, such as changes in emergency visits for asthma among children with changes in 1-hour ozone concentrations.[58] In addition to dose–response

[57] UN-ECE, 1979 *Convention on Long-range Transboundary Air Pollution and its Protocols.* ECE/EB.AIR/50, United Nations Economic Commission for Europe, Geneva, Switzerland, 1996.

[58] I. Romieu and M. Hernandez, Air pollution and health in developing countries: review of epidemiological evidence, in *Health and Air Pollution in Developing Countries*, ed. G. McGranahan and F. Murray, Stockholm Environment Institute, York, 1999, pp. 43–56.

[59] M. L. Cropper, N. B. Simon, A. Alberini and P. K. Sharma, *The Health Effects of Air Pollution in Delhi, India*, The World Bank, PRD Working Paper 1860, 1997.

Relative grain yield (%)
AOT40 (ppm.h)
y=99.5 - 1.7x
r²=0.88

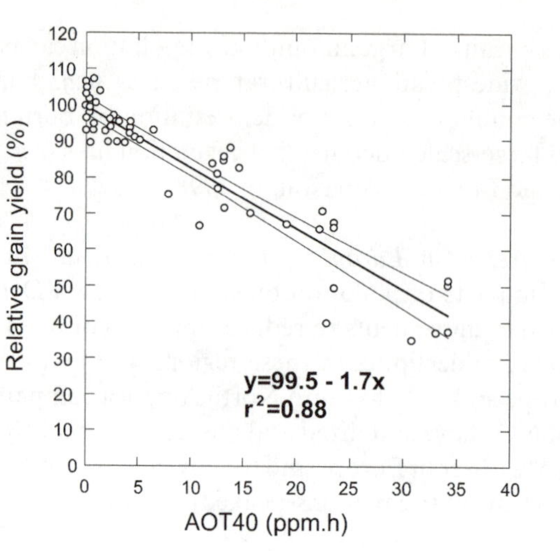

Figure 13 The relationship between relative grain yield of Spring wheat and ozone exposure expressed as an AOT40 index grown in field experiments at a range of locations across Europe[63]

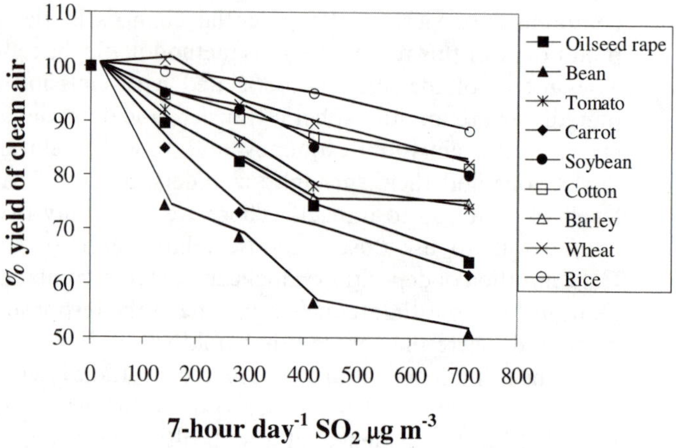

% yield of clean air
7-hour day⁻¹ SO₂ µg m⁻³
Oilseed rape
Bean
Tomato
Carrot
Soybean
Cotton
Barley
Wheat
Rice

Figure 14 Percentage yields of crop plants exposed to different concentrations of SO_2 in comparison with plants in clean air. % yield of clean air = yield of plant exposed to SO_2/yield of plant in clean air (drawn from data in ref. 64)

relationships, air quality guidelines are frequently used to assess whether pollutant concentrations are damaging to human health.

For corrosion of materials, Kucera *et al.*[60] have developed methods to estimate the degree of corrosion to cities and estimate the economic consequences of pollution increases or reductions. Essentially they divided cities into pollution strata based upon SO_2 concentrations. In each stratum, an inventory of stock at risk is made and linked to estimated change in service life based upon

[60] V. Kucera, J. Henriksen, D. Knotkova and C. Sjöström, Model for calculations of corrosion costs caused by air pollution and its applications in three cities, *Proceedings 10th European Corrosion Congress, Barcelona,* Swedish Corrosion Institute, Stockholm, 1997.

[61] M. R. Ashmore and R. B. Wilson, *Critical Levels of Air Pollutants for Europe*, Department of the Environment, London, 1993.

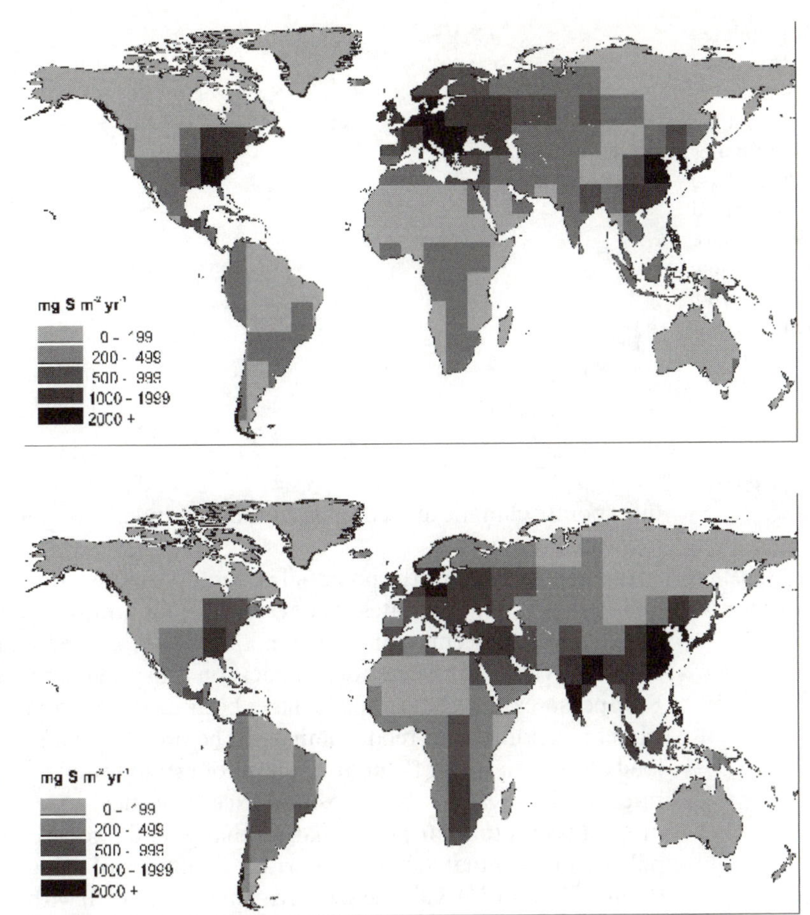

Figure 15 Annual total (wet plus dry) deposition of sulfur in 1990[65,41] calculated by the MOGUNTIA Model (at 10 × 10 degree resolution) using GEIA[66] emission estimates for 1990 (mg S m^{-2} yr^{-1})

Figure 16 Annual total (wet plus dry) deposition of sulfur in 2050[67,41] calculated by the MOGUNTIA Model (at 10 × 10 degree resolution) using an emission projection for 2050 derived from an IPCC SRES A2 scenario[68] (mg S m^{-2} yr^{-1})

dose–response relationships for different materials. Using maintenance and replacement costs the economic consequences of emission reductions to UN/ECE Protocol levels were calculated for cities in Europe. Total cost savings were estimated at US$ 9 billion.

On the basis of the exposure–response experiments, so-called ‘critical levels’ have been established which represent values above which responses, usually detrimental, have been observed.[61] These can be compared to concentration estimates to determine in which areas gaseous pollutants may be having an impact on crops or forests. Dose–response relationships also exist for ozone impacts on certain crop and tree species (*e.g.* Figure 13 for wheat) and for SO_2 concentrations with crop species (*e.g.* Chinese data in Figure 14). The ozone relationship with an accumulated dose above 40 ppb (AOT40) is not sufficient for crop yield loss estimations as the impact in the field is more closely related to absorbed dose into the stomata of plants, rather than concentrations in air. Efforts are being made to develop flux–response relationships which will

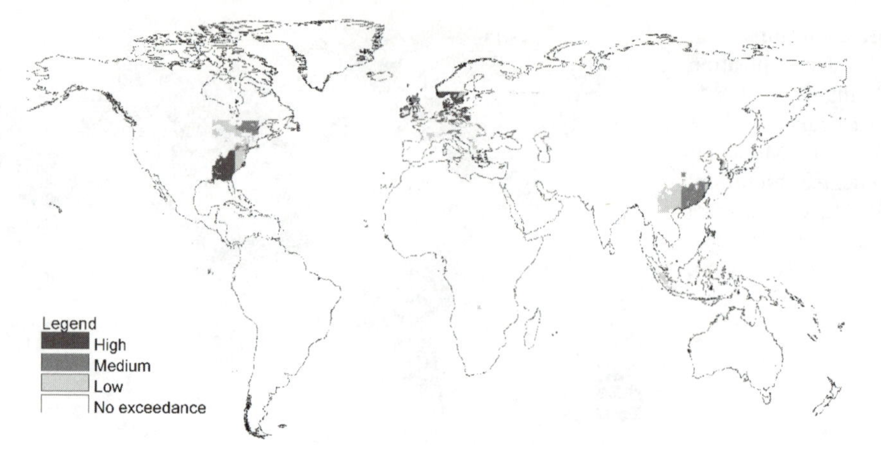

Figure 17 Estimated exceedance of critical loads (applied to the sensitivity map – Figure 8) by 1990 sulfur deposition estimates minus neutralizing base cation deposition derived from soil dust. (Exceedance scale in meq m^{-2} yr^{-1}: low 0–50; medium 50–100; high >100)[41]

incorporate climatic and edaphic conditions and allow more accurate crop yield loss estimations.[62]

In order to examine the potential impacts of acidification, it is possible to use S and N deposition estimates, as developed from atmospheric transfer models in combination with emission estimates and scenarios (see Figures 15 and 16), with threshold values for deleterious impacts on ecosystem structure and function. In Europe, so-called 'critical loads' have been developed that represent threshold values to acidification-related impacts. The simplest methods to estimate critical loads has been to set them at the level of estimates of soil mineral weathering rates.[67] The areas where deposition exceeds critical loads can be mapped and prove a useful tool for policy makers that would like to consider the impact of policies on eventual acidification risks. Examples of using the deposition maps (Figures 15 and 16) with the sensitivity map (Figure 8) with critical loads applied (and taking into account estimates for the neutralizing base cation deposition) are shown in Figures 17 and 18. The areas exceeded represent areas worthy of further study as they may be at risk from acidification-related impacts. It can be seen that, in addition to areas where acidification is well known in Europe and North America, there are widespread areas in Asia and also more limited areas in southern Africa which implies that these areas could be at risk from acidification-related impacts.

[62] L. D. Emberson, M. R. Ashmore, H. M. Cambridge, D. Simpson and J.-P. Tuovinen, Modelling stomatal ozone flux across Europe, *Environ. Pollut.*, 2000, **109**, 403–413.

[63] J. Fuhrer, L. Skärby and M. R. Ashmore, Critical levels for ozone effects on vegetation in Europe. *Environ. Pollut.*, 1997, **53**, 365–376.

[64] Z. Feng, H. Cao and S. Zhou, *Effects of Acid Deposition on Ecosystems and Recovery Study of Acid Deposition Damaged Forest*, China Environmental Science Press, Beijing, 1999 (in Chinese).

[65] J. Langner and H. Rodhe, A global three-dimensional model of the tropospheric sulphur cycle. *J. Atmos. Chem.*, 1991, **13**, 225–263.

[66] C. M. Benkowitz, M. T. Scholtz, J. Pacyna, L. Tarrason, J. Dignon, E. C. Volder, P. A. Spiro, J. A. Logan and T. E. Graedel, Global gridded inventories of anthropogenic emissions of sulphur and nitrogen, *J. Geophy. Res.*, 1996, **101** (D22), 29 239–29 253.

[67] J. Nilsson (ed.), *Critical Loads for Nitrogen and Sulphur*, Miljörapport 1986:11, Nordic Council of Ministers, Copenhagen, 1986.

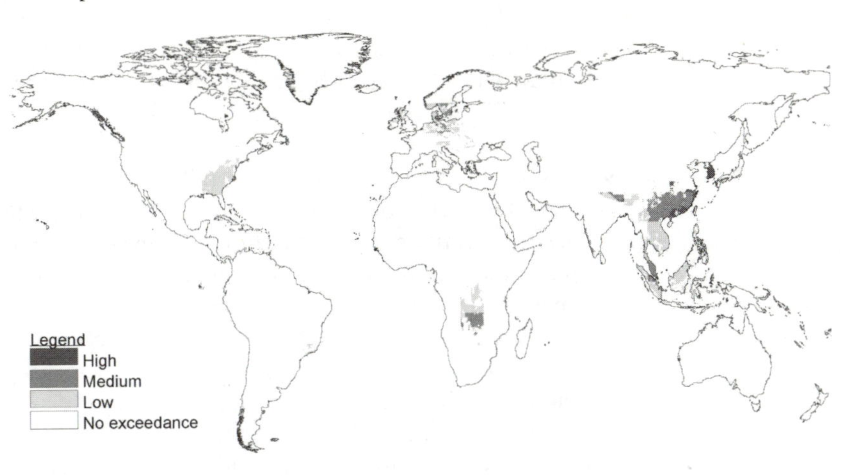

Figure 18 Estimated exceedance of critical loads (applied to the sensitivity map – Figure 8) by 2050 sulfur deposition estimates minus neutralizing base cation deposition derived from soil dust. (Exceedance scale in meq m^{-2} yr^{-1}: low 0–50; medium 50–100; high >100)[41]

4 Policy Development for Air Pollution Abatement

National and Regional Action

The smogs in Europe and North America in the 1950s triggered the development of environmental policies and measures to combat air pollution issues. In recent years, air pollution abatement strategies and environmental laws and institutions have developed in most countries and in regions such as the area covered by the UN Economic Commission for Europe (UN/ECE) and the European Union. Command and control policy, *via* direct regulation, has been the most prominent policy instrument. Traditionally, regulation is carried out by sector but there is a growing realization that an integrated approach is required. Lack of manpower, the methods of implementation and control, and the level of institutional co-ordination and policy integration can hamper this approach.[45] This is particularly true of efforts to regulate air pollution in developing countries.

In recent years, there has been a shift away from regulation and an increased use of economic instruments, tax and subsidy reform, reliance on voluntary action by the private sector, accompanied by more public and NGO participation (UNEP, 2000).[45] This has resulted in more emphasis on attempts to move away from end-of-pipe measures to preventive measures including a lowering of the intensity of energy use by increases in energy-use efficiency and a switch away from non-renewable fossil fuels to renewable energy resources. These developments are fuelled by the increasing complexity of environmental regulation and high control costs as well as demands from the private sector for more flexibility, self-regulation and cost-effectiveness.

International Action

In the last two decades the development of policy on air pollution issues has been greatly assisted by Multi-lateral Environmental Agreements (MEAs). MEAs have proven to be a powerful tool for tackling air pollution problems, especially from a transboundary perspective (see below). Although most MEAs are relatively weak (many signatory countries are not obliged to take any more action than what has already been anticipated) their many benefits, some of which are often overlooked, can be summed up as follows:[69]

- Even seemingly weak demands entail that the more reticent countries take some action that otherwise would not have been taken;
- The preparations for negotiations involve the international exchange of large amounts of information, as well as interaction between national experts, scientists and policy-makers;
- The process of elaborating international agreements entails the generation of new data which can then be used to increase public awareness and opinion, both nationally and internationally;
- The negotiations attract media attention which helps to raise the profile of the problem and in turn generates public awareness and knowledge.

Disadvantages can be that extensive and time-consuming research activities and the potentially lengthy negotiation procedures (*e.g.* the ratification process) can be used as an excuse for not taking action. However, the activities underpinning negotiation may directly or indirectly impact significantly on national decision-making when it comes to taking measures for reducing or preventing emissions. Indeed, the process of arriving at an agreement can be considered as contributing as much, or more, to the protection of the environment, as the agreement itself.

In addition to binding MEAs, progress in environmental protection can be promoted by the non-binding agreements, such as Local Agenda 21, and environmental clauses or principles in wider agreements, such as regional trade treaties.

The Status of Air Pollution Policy in Different Regions of the Globe

Europe. Excess deaths caused by urban smogs in the UK led to several policy measures outlined in the Clean Air Acts of 1956 and 1968. Smokeless zones were introduced and industry was moved away from urban centres and was required to install taller stacks. The development of tall stacks and other measures in Europe, coupled with the dramatic rate of industrialization and fossil fuel use, caused the development of regional problems such as ecosystem acidification.

In the 1960s and 1970s it was realized that acidification was due, in part, to the transboundary transfer of air pollution which entailed that individual countries could not solve their problems alone. This, then, necessitated the development of international agreements for co-ordinated emission control and eventually, in 1979, the international policy negotiations of the United Nations Economic Commission for Europe (UN/ECE) led to the development of the Convention on Long-range Transboundary Air Pollution (LRTAP). The LRTAP Convention

[68] http://sres.ciesin.org

[69] C. Ågren, The European example of emission control, in *Regional Air Pollution in Developing Countries.* Background Document for Policy Dialogue, Stockholm Environment Institute, York, 1998, pp. 119–126.

was the first MEA concerning air pollutants and has been very successful, resulting in Protocols for the control of sulfur, nitrogen and volatile organic compounds emissions.[70] More recently, in December 1999, the UN-ECE 'Multi-Pollutant, Multi-Effect' Protocol was signed.[71] The Protocol simultaneously addresses acidification, eutrophication and ground-level ozone, which continue to have serious effects in the UN/ECE region. Furthermore, in tackling ammonia emissions from agriculture, it is the first Protocol to the Convention to address the agricultural sector.

The first success of the LRTAP Convention, in 1985, was the First Sulfur Protocol which required signatories to reduce their national emissions of sulfur, or their transboundary fluxes, by at least 30% as quickly as possible and at the latest by 1993, using 1980 as the base year. The approach of an equal percentage of reduction for every country proved unpopular with some countries and difficult to justify on scientific grounds. Thus efforts were made to develop an 'effects-based approach' whereby reductions of emissions would be negotiated on the basis of the effects of air pollutant loads. This led to the development of the 'Critical Loads' approach which was initially defined as 'the highest load [of pollutant] that will not cause chemical changes leading to long-term harmful effects on the most sensitive ecological systems'.[65] The concept was first used in Canada and Sweden in the early 1980s to indicate deposition values below which there would be no risk to lake ecosystems of differential sensitivity to acidification.[72,73] Canada used these 'Target Loads' in its negotiations with the USA. In Europe the first outcome of the critical load approach was the Second Sulfur Protocol, which was signed in 1994. The approach established differing requirements for each country, the aim being to attain the greatest effect for the environment at least overall cost. The focus of reduction strategy had therefore changed from a focus on emission standards to the maintenance of a targeted air quality.

In addition to the protocols of the LRTAP Convention, various EU Directives and strategies (including the 'Large Combustion Plant Directive' and the 'Community Strategy to Combat Acidification'[74]) have also been developed. EU legislation has the advantage that it is more binding than the Protocols of the LRTAP Convention. EU strategies have also been based on effects-based approaches, rather than being based solely on technological or economic feasibility.

[70] UN/ECE, *Convention on Long-range Transboundary Air Pollution and its Protocols*. ECE/EB.AIR/50, United Nations Economic Commission for Europe, Geneva, Switzerland, 1996.

[71] UN/ECE, *Protocol to Abate Acidification, Eutrophication and Ground-level Ozone*. United Nations Economic Commission for Europe, Geneva, Switzerland, 2000.

[72] F.C. Elder and T. Brydges, *Effects of Acid Precipitation on Aquatic Regimes in Canada*, The Canadian contribution to the Committee on the Challenges of Modern Society. NATO Panel 3 Report, 1983.

[73] W. Dickson, Some data on critical loads for nitrogen and sulphur deposition, in *Critical Loads for Nitrogen and Sulphur*, ed. J. Nilsson, Miljörapport 1986:11, Nordic Council of Ministers, Copenhagen, 1986, pp. 143–158.

[74] D. Gillies, *A Guide to EC Environmental Law*, Earthscan Publications Ltd., London, 1999.

North America. In the USA the smog problem experienced in Los Angeles and other major cities led to the enactment of the first federal air pollution legislation in 1955, which provided federal support for air pollution research, training and technical assistance. In 1970 the US Clean Air Act was passed and, whereas the UK Clear Air Act of 1956 tackled smoke pollution, the US Act dealt, in addition, with pollutants such as ozone that contribute to photochemical smog. As these were more difficult to control, the non-attainment of set standards was so widespread that the Act was perceived as a failure. However, this situation was improved by the extensive amendments that resulted in the Clean Air Act 1990.[75] Areas of non-attainment required car-pooling, clean fuel and auto inspection and maintenance programmes. The need for the introduction of new types of environmental policy became apparent and has resulted in the development of market based policies, such as tradable emission permits and agricultural subsidy reform. Voluntary policies and private sector initiatives, often in combination with civil society, are also increasingly used for pollution reduction. Furthermore, the value of participation of different stakeholders in the development of environmental policy instruments, such as the public, NGOs, industry *etc.*, is now being realized. In 1980 the Acid Precipitation Act first authorized a cooperative federal program, the National Acid Precipitation Assessment Program (NAPAP), to coordinate acid rain research and report the findings to Congress. The research, monitoring, and assessment efforts by NAPAP and others in the 1980s culminated in Title IV of the 1990 Clean Air Act Amendments, also known as the Acid Deposition Control Program.

In 1969, the federal government of Canada introduced the Canadian Clean Air Act to address and monitor the problem of air pollution. At the present time the emphasis is on regulatory reform, federal /provincial policy harmonization and voluntary initiatives. In order to study regional patterns of air pollution The Canadian Air and Precipitation Monitoring Network (CAPMoN) was set up.

In 1991, after initial failures in the 1980s,[76] bilateral negotiations between Canada and the US resulted in the Air Quality Agreement to address transboundary air pollution.[77] Acid rain was the initial focus of co-operative transboundary efforts under the Air Quality Agreement, but, after significant progress, Canadian and US environmental ministers signed a Joint Plan of Action in 1997 for Addressing Transboundary Air Pollution on ground-level ozone and particulate matter. In December 2000 co-operation between the two countries resulted in the signing of an ozone annex to the Air Quality Agreement, addressing their common concerns about ground-level ozone's transboundary impacts. Concurrently, both governments are undertaking co-operative efforts in particulate matter modelling, monitoring, and data analyses to assess transboundary PM impacts and support development of a joint action plan to address the issue.

[75] R. W. Boubel, D. L. Fox, D. B. Turner and A. C. Stern, *Fundamentals of Air Pollution*, Third Edition, Academic Press, 1994.

[76] J. Schmandt and R. Hilliard, eds., *Acid Rain and Friendly Neighbors: The Policy Dispute between Canada and the United States*, Duke University Press, Durham, 1985.

[77] *Air Quality Agreement between Canada and the US – Progress Report 2000*. International Joint Commission, Ottawa and Washington.

Asia. In general, widespread concern over pollution and natural resources has led to legislation to curb emissions and conserve natural resources.[45] Governments are increasingly active in promoting environmental compliance and enforcement, although the latter is still a problem in parts of the region. The use of economic incentives and disincentives is becoming more widespread as a method of achieving environmental protection and for promoting resource use efficiency and pollution fines are common. Industrial groups in both low- and high-income countries are becoming more receptive to environmental concerns over industrial production. Environmental funds have been established in many countries and have contributed to the prominent role of NGOs in environmental action. Many countries support public participation and in some countries this is a requirement of law. However, education and awareness levels amongst the public are often low, and the environmental information base in the region is weak. Multilateral agreements in three major regions are beginning to take shape: in South Asia, South-East Asia and North-East Asia. This has been prompted by concern over transboundary air pollution, that was graphically demonstrated in South-East Asia during the Indonesian forest fires, but which has been demonstrated by monitoring programmes in China, Korea and Japan, and to a lesser extent in other Asian countries.

South Asia. Of the Asian sub-regions, South Asia (including Bangladesh, India, Pakistan, Nepal, Iran, Bhutan, Maldives and Sri Lanka) is at the earliest stage of the air pollution policy cycle but has started to make progress in recent years. The most significant development in recent years is the 'Malé Declaration on Control and Prevention of Air Pollution and its likely Transboundary Effects for South Asia', which was adopted at the 7th Governing Council meeting of SACEP (South Asia Co-operative Environment Programme) where Environmental Ministers met in Malé, Maldives, in 1998. The Declaration encourages intergovernmental co-operation to address the increasing threat of transboundary air pollution and its associated impacts. At the heart of the Declaration is the promotion of an institutional framework that links scientific research and policy formulation. The first phase of the implementation of the Declaration has now been completed and resulted in baseline reports of the status of air pollution in the region and the development of national and sub-regional action plans. The second phase will continue to promote regional co-operation and will see the establishment of a standardized monitoring network in the region.

South-East Asia. In South-East Asia the issue of transboundary pollution was first highlighted in the 1990 Kuala Lumpur Accord on Environment and Development. The 1992 Singapore Summit identified it as among the major environmental concerns of the Association of South-East Asian Nations (ASEAN). In 1995 the ASEAN Co-operation Plan on Transboundary Pollution[78] was adopted. The ASEAN Co-operative Plan on Transboundary Pollution consists of three programme areas, namely: transboundary atmospheric pollution; transboundary movement of hazardous waste; and transboundary shipborne pollution. The programme area on transboundary atmospheric pollution has been set up:

[78] ASEAN, *ASEAN Co-operative Plan on Transboundary Pollution.* ASEAN Secretariat, Jakarta, 1995.

- to assess the origin and cause, nature and extent of local and regional haze incidents;
- to prevent and control the sources of haze at both national and regional levels by applying environmentally sound technologies and by strengthening both national and regional capabilities in the assessment, mitigation and management of haze;
- to develop and implement national and regional emergency response plans.

The Haze Technical Task Force was established in 1995 to operationalize and implement the measures included in the ASEAN co-operation plan on transboundary pollution. Following the 1997 fire-and-haze event, a Regional Haze Action Plan was formulated by the Haze Technical Task Force and was endorsed by the ASEAN Ministerial Meeting in 1997. The Regional Haze Action Plan has an operational focus, intending to identify specific actions to be taken at regional, sub-regional and national levels to prevent man-induced haze events. South-East Asian countries are also included in the EANET Programme described below.

North-East Asia. Concern regarding ongoing environmental degradation in the North-East Asian region, resulting from the dynamic development of the region, despite individual nations' efforts to prevent environmental pollution led to the initiation of the Tripartite Environment Ministers Meetings (TEMM)[79] amongst Korea, China and Japan. In their first communiqué in 1999 the ministers recognized that China, Japan, and Korea are playing important roles in economic and environmental cooperation in the North-East Asian region and that close cooperation among the three nations is indispensable to sustainable development in North-East Asia. The ministers shared the view that the following areas of cooperation should be given priority: raising awareness that the three countries are in the same environmental community; activating information exchange; strengthening cooperation in environmental research; fostering cooperation in the field of environmental industry and on environmental technology; pursuing appropriate measures to prevent air pollution and to protect the marine environment; and strengthening cooperation on addressing global environmental issues, such as biodiversity and climate change. They also expressed their intention to cooperate to tackle these issues. The awareness of air pollution issues in the region in the last decade has been aided by the development of several regional multilateral initiatives including:

(i) *Joint Research Project on Long-Range Transboundary Air Pollutants in Northeast Asia (LTP project)*

This joint project was started in 1995 by the National Institute of Environmental Research (NIER) of the Republic of Korea (ROK) to investigate the current situation of transboundary movement of air pollutants and to support joint research amongst ROK, China and Japan on long range transboundary air pollutants in North-East Asia.

(ii) *Acid Deposition Monitoring Network in East Asia (EANET)*

[79] Communiqués of Tripartite Environment Ministers Meetings amongst China, Japan and Korea.

The objective of this network is to carry out monitoring of acid deposition by harmonized methodologies in East Asia, and thus to create a common understanding of the state of acid deposition which will be the scientific basis for further steps such as measures to reduce adverse impacts on the environment caused by acid deposition. In 1998, the network started its preparatory phase activities. The Interim Network Center was established in Niigata, Japan. The network was participated by 10 countries: China, Indonesia, Japan, Malaysia, Mongolia, Philippines, Republic of Korea, Russia, Thailand and Vietnam. The Second Intergovernmental Meeting for EANET concluded that the preparatory phase activities were successful in demonstrating the feasibility of EANET and decided, through the joint announcement, that the participating countries would cooperatively start the activities of EANET on a regular basis from January 2001. The Ministry of the Environment, Japan, would serve as the Interim Secretariat for EANET until UNEP assumed the role of the Secretariat. Acid Deposition and Oxidant Research Center (ADORC), which is located in Niigata, Japan, was designated as the Network Center for EANET, and started its Network Center functions from January 2001.

(iii) UN ESCAP Expert Group Meeting on Emissions Monitoring and Estimation[80]
A regional project on 'Technical Assistance for Environmental Cooperation in North East Asia' was initiated in 1996 by the UN Economic and Social Commission for Asia and the Pacific (ESCAP) with the support of the Asian Development Bank (ADB). The project focuses on promoting regional cooperation for environmental protection amongst China, the DPRK, Japan, Mongolia, ROK and the Russian Federation, and has included technology demonstrations and workshops, expert meetings and review meetings.

Africa. Many African countries are implementing new national and multilateral environmental policies; however, their effectiveness is often low due to the lack of staff, expertise, funds and equipment for implementation and enforcement. Current environmental policies are mainly based on regulatory instruments but some countries have begun to consider a broader range, including economic incentives implemented through different tax systems. Some industries have adopted cleaner production options but this is not widespread in the region; however, some multilateral corporations, large-scale mining companies and local enterprises have voluntarily adopted precautionary environmental standards in recent times.

Southern Africa. While many of the Southern African Development Community (SADC) countries acknowledge that air pollution is a problem, only South Africa and Zambia are seen as the major contributors to industrial air pollution. Practices such as domestic fuel burning and grassland burning for grazing are widespread in the region, but the impacts on human and ecosystem health are not apparent. In terms of atmospheric pollution, the main issues in policy initiatives are on ozone depleting substances and greenhouse gases (driven by the global process). Virtually all SADC member states have policies on air pollution control

[80] Environment Agency of Japan, *Summary of the Expert Group Meeting on Emissions Monitoring and Estimation, 27–29 January, 1999*, Niigata, Japan, 1999.

in one form or another. These have either been covered in National Conservation Strategies (NCS), National Environmental Action Plans (NEAPs) and/or a new generation of environmental laws. While the necessary laws and policies may be in place, their implementation is lacking, with the result that there is little or no regulation of air pollution sources.

The majority of SADC countries have ratified both the 1987 Montreal Protocol on Substances that Deplete the Ozone Layer and the 1992 Framework Convention on Climate Change. While this shows the commitment by SADC countries to global issues, the immediate problems of air pollution do not seem to have had similar attention. Action plans exist for the phase-out of ozone-depleting substances and reduction of greenhouse gases but no such plans exist for air pollution in general.

Latin America. Air pollution problems in Latin America are not as widespread as in Asia and, as a consequence, are not as high on political agendas, except in the 'megacities'. Some of these cities are amongst the most polluted in the World. In urban environments air pollution has been considered a serious problem for many years[81] and emissions of sulfur dioxide, nitrogen oxides, ammonium and, in particular, concentrations of tropospheric ozone, have been increasing over the last few decades.[82] Only when large point sources of pollution, such as power stations, are situated near national boundaries does the pollution result in a transboundary problem. However, as emissions increase the problem will become increasingly regional.

The air pollution issue has been addressed at the local scale in cities, such as Mexico City, where certain air pollution control measures have been implemented imposing restricted use of motor vehicles, temporary closure of industries and policies to move industries away from cities. Until recently, regional initiatives in South America have been limited to activities associated with the climate change debate and the Kyoto Protocol, but there are now activities occurring with regard to regional air pollution.

In South America, the countries of the MERCOSUR Free Trade Agreement (Argentina, Brazil, Uruguay and Paraguay) regard air pollution not as a transboundary problem, but more as a shared problem in the region.[83] The focus is more on trade issues and on the harmonization of legal frameworks to ensure that there is a level playing field for industry with respect to environmental regulations.

[81] H. Rodhe, E. Cowling, I. Galbally, J. Galloway and R. Herrera, Acidification and regional air pollution in the tropics, in H. Rodhe and R. Herrera, eds., *Acidification in Tropical Countries*, SCOPE 36. John Wiley & Sons, Chichester, 1988.

[82] E. Sanhueza, Deposition in South and Central America, in *Global Acid Deposition Assessment*, D. M. Whelpdale and M. S. Kaiser, eds., WMO, Global Atmospheric Watch Publ. No. 106, 1997, Ch. 9, pp. 135–144.

[83] W. K. Hicks, J. C. I. Kuylenstierna, V. Mathur, S. Mazzucchelli, M. Iyngararasan, S. Shrestha and A. M. van Tienhoven, Development of the regional policy process for air pollution in Asia, Africa and Latin America, in *Air Pollution and the Forests of Developing and Rapidly Industrializing Countries*, eds. J. L. Innes and A. H. Haron, No. 5 in the IUFRO Research Series, 2000.

5 Initiatives to Improve Urban Air Quality in Developing Countries

There has been a steady improvement in urban air quality in industrialized countries, including cities where increasing traffic is creating a different set of problems. Thus, many have hoped that cities in the developing world would be able to side-step some of the more drastic urban air pollution problems that had been experienced in Europe, North America and Japan. Such optimism has not been well-founded; many of the rapidly growing cities of the developing world have followed the same path of air pollution build-up as that experienced by developed county cities in earlier decades. For example, an assessment of air quality management in Ankara by the United Nations Environment Programme[84] has concluded:

> 'The causes of air quality problems in Ankara, and the adopted solutions, are similar to those of London in the 1960s. Air quality problems are predominantly caused by domestic burning of low-grade coal and are particularly acute during winter in weather characterized by low wind speeds and the formation of temperature inversions. These factors combine to cause high concentrations of particulate matter and SO_2 (and its secondary products). Furthermore, in London in the 1960s and currently in Ankara, problems are, for the most part, being remedied by increasing the domestic use of cleaner fuels.'

It is disappointing that there was, after almost 40 years, such a close rerun of the episodes experienced, at some cost (see Section 1), elsewhere. However, these disappointments have probably led to some of the situations where concerted efforts are now being made to improve urban air quality.

The Kathmandu Valley

Air quality in the Kathmandu Valley, the largest urban conurbation in Nepal, is acknowledged to have deteriorated.[85] Fuelwood, agricultural residues and animal waste make up the bulk of fuel consumption. The number of vehicles in the Valley is rising rapidly. A large number of industries are classified as 'polluting'. It is not surprising therefore that good visibility is down, on average, to five days in a month in Kathmandu, respiratory diseases have increased in recent years and low crop yields are recorded as being related to high levels of dust deposition. The Nepalese Government has introduced measures to deal with industry- and transport-generated pollution including the restriction of the registration of two-stroke engine vehicles in important tourist centres (as these account for over 50% of vehicles). There is a recent ban on three-wheeler tempos in the Kathmandu Valley. Now there are up to 450 electric-powered *Safa* tempos in use.

[84] MARC, *Air Quality Management and Assessment Capabilities in 20 Major Cities*, published on behalf of UNEP and WHO by the Monitoring and Assessment Research Centre, London, 1996.

[85] UNEP, *Nepal: State of the Environment 2001*, United Nations Environment Programme, Pathumthani, Thailand, 2001.

Delhi

Many Indian towns and cities experience air pollution problems similar to the Kathmandu Valley. In Delhi (as well as in other areas), there has been a rather different strategy, focused on the industrial sector as well as transport. Recently (in November 2000) polluting industries in residential areas have been the subject of relocation orders, into districts designated for industry. Some existing industrial units have been 'sealed'. Not surprisingly, this has resulted in considerable resistance from industry who claim the sites for relocation lack adequate infrastructural provision. This emphasizes the need to deal with such problems in an integrated manner.

Bogota

In this seven-to-eight million strong, high altitude, capital city of Colombia, transportation is rapidly becoming a nightmare. There is now a high incidence of respiratory infection in the population and travel times to work are commonly in excess of two hours. An alternative transport system for the city is planned, and part already under construction, that is aimed at the needs of 98% of the population. The plan relies upon:

- an aggressive programme of parking control;
- even/odd car restraint days ('Pico y Placa');
- a pedestrian zone and pavement plan that is already under construction;
- a 'world-level' bicycle transportation scheme for which the infrastructure is already under construction;
- 'car-free' days in the city;
- the replacement of the planned metro line by a high-capacity fine-mesh 'Transmilennio' system made up of articulated buses;
- a vigorous policy aimed at fleet renewal of existing vehicles.

Many problems remain. There are more than 30 000 buses, busetas, minibuses and collectivos, and over 55 000 taxis. The condition of many of these vehicles means that emissions to the atmosphere, noise and accidents are an enormous problem. But if a developing country city like Bogota is able to find a way out of such problems and devise sustainable solutions, then it will prove a beacon for other cities in the developing world to tackle similar problems.

Hong Kong

Hong Kong has no indigenous fuel. As well as importing oil, diesel and liquid petroleum gas, it imported 10 Mt of coal per annum, over 60% of which was used for electricity generation. Coal-fired power generation contributed significantly to emissions of particulate matter, SO_2 and NO_x. However, Hong Kong has a developing environmental protection system and the Air Pollution Control Ordinance (1990) was a means of controlling emissions from power plants. Electrostatic precipitators have effected some control of particulates and the use of low sulfur coal, together with flue gas desulfurization units on some power plants has dramatically reduced SO_2 emissions. A 50% reduction of NO_x emissions has been achieved on units with low NO_x burners fitted. Since 1991, emission standards for coal-fired power plants have been revised downwards from 125 to 50 $mg\,m^{-3}$, 2100 to 200 $mg\,m^{-3}$ and 2200 to 670 $mg\,m^{-3}$ for particulate matter, SO_2 and NO_x, respectively. In some polluted districts there was a rapid decrease in ambient concentrations of SO_2 by 80%.

There are still air quality problems in Hong Kong, however. Both respirable suspended particulates and NO_x exceed the limits established. Polluted districts of Hong Kong, in a study to assess the effect of the emission controls,[86] had worse respiratory symptoms and bronchial conditions in children than the less polluted districts, but it was in the former districts that the greatest improvement was observed. In adults, an epidemiological study demonstrated that there was a strong, positive correlation between pollutant levels and daily hospital admissions for cardiovascular and respiratory conditions.

It can be concluded that air pollution controls do bring significant benefits in terms of health improvement and a reduction in premature deaths, particularly in the elderly. Moreover, the response time for measurable health benefits to be realized is short – weeks or months rather than years.

6 Future Action

As urban centres continue their rapid growth in developing countries, air pollution problems are likely to increase in their frequency and severity. The initiatives being taken in 'megacities' in certain of these countries to begin to deal with air quality problems, by regulation, structural change and the search for a more sustainable system, will act as lessons and, where successful, models for more widespread action. Progress is slow but the wealth of experience that exists worldwide is an invaluable resource for concerted and dedicated action.

In addition to the actions at urban scale there are many activities on-going in Asia, Africa and Latin America at national and international scales. For example, the Chinese government have been introducing abatement strategies and developing important policy instruments.[87] Policies have been implemented to increase efficiency, reduce sulfur contents in coal and also relatively expensive Flue-Gas-Desulfurization (FGD) equipment has been installed in a number of power plants in China.[88] FGD has also been in stalled in countries such as Thailand[89] and are planned/being constructed in other South-East Asian countries. Particulate emission reduction technologies (especially Electrostatic

[86] G. McGranahan and F. Murray (eds.), *Health and Air Pollution in Rapidly Developing Countries*, Stockholm Environment Institute, Stockholm, 1999.

[87] H. M. Seip, P. Aagaard, V. Angell, O. Eilertsen, T. Larssen, E. Lydersen, J. Mulder, I. P. Muniz, A. Semb, T. Dagang, R. D. Vogt, Xiao Jinshong, Xiong Jiling, Zhao Dawei and Kong Guohui, Acidification in China: assessment based on studies at forested sites from Chioongqing to Guangzhou, *Ambio*, 1999, **28**, 522–528.

[88] H. N. Soud and Z. Wu, *East Asia – Air Pollution Control and Coal-Fired Power Generation*, IEA Coal Research, London, 1998.

[89] H. N. Soud, *Southeast Asia – Air Pollution Control and Coal-Fired Power Generation*, IEA Coal Research, London, 1997.

Precipitators) are becoming widespread in Asia.[88,89] In addition to national policies, many regional, inter-governmental initiatives have started in different parts of the World. The experience from Europe through the UN/ECE LRTAP Convention has been that such international cooperation can galvanize countries of that region into taking action. Hopefully this will be the case in Asia and Africa, even though these regions differ greatly from Europe in many respects.

Influence of Climate Variability and Change on the Structure, Dynamics and Exploitation of Marine Ecosystems

MANUEL BARANGE

1 Introduction: The Global Ocean's Response to Climate Variability and Change

Global warming has been the subject of considerable debate in recent years. Present estimates indicate that the Earth's surface temperature has increased at a rate of 0.1 °C per decade over the 20th century (Figure 1).[1] Although some of this warming may be caused by natural variability there is now consistent evidence of an anthropogenic signal in the climate record for at least the last 50 years, but possibly over the last 100 years.[2] Such a signal should be expected. In the last century the world's population has increased by a factor of four, and industrial output has increased 40 times.[3] Human action has transformed nearly 50% of the land surface, and the burning of fossil fuels has increased the atmospheric concentration of carbon dioxide (CO_2) to levels not exceeded in at least 420 000 years.[4] These are some of the environmental costs for the planet of improving the standards of living, education and leisure of about a quarter of the world's population. Can we achieve equal successes for the other 75% of the world's population without destroying our support systems? Based on IPCC scenarios of future emissions, temperature increases of between 1.4 and 5.8 °C between 1990 and 2100 are expected.[2] A further 1% annual increase in CO_2 will double the current CO_2 levels in the atmosphere in 70 years, and quadruple them in 140 years.[5] What can we expect from these changes? We know that the Earth does not respond to global changes uniformly, and that the complexity of its response sometimes defies purely mechanical analysis (Table 1). Its vital organs, switches and choke points respond to global changes in non-linear, asynchronous ways, resulting in positive as well as negative feedbacks. The mechanisms at play have

[1] M. E. Mann, R. S. Bradley and M. K. Hughes, *Geophys. Res. Lett.*, 1999, **26**, 759–762.
[2] IPCC, Third assessment Report of working group 1. Summary for policymakers, 2001.
[3] J. R. McNeill, *Something New Under the Sun: An Environmental History of the Twentieth-Century World*, Allen Lane/W. W. Norton, 2000.
[4] J. R. Petit, J. Jouzel, D. Raynaud, N. I. Barkov, J.-M. Barnola, I. Basile, M. Bender, J. Chappellaz, M. Davis, G. Delaygue, M. Delmotte, V. M. Kotlyakov, M. Legrand, V. Y. Lipenkov, C. Lorius, L. Pepin, C. Ritz, E. Saltzmank and M. Stievenard, *Nature*, 1999, **399**, 429–436.
[5] S. Manabe and R. J. Stouffer, *Nature*, 1993, **364**, 215–218.

Issues in Environmental Science and Technology, No. 17
Global Environmental Change

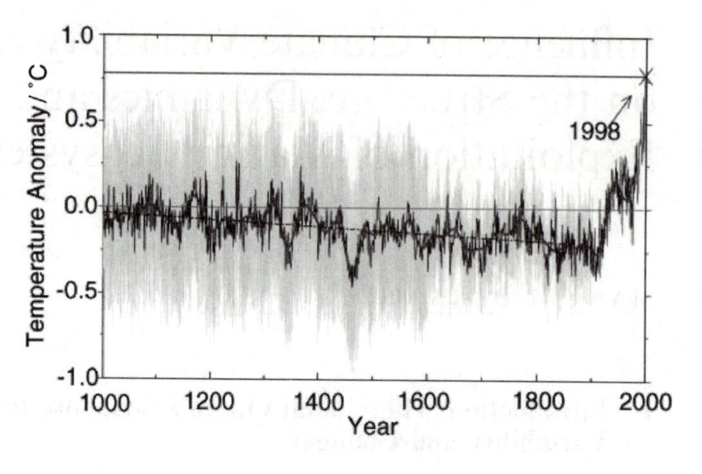

Figure 1 Mean annual temperature variations over the northern hemisphere for the last 1000 years[1]

Table 1 Climate scenarios for the 20th and 21st centuries[2]

Observations for the 20th century

- The global average surface temperature has increased by 0.6 °C
- Greater increases have been observed in night-time *versus* daytime temperatures
- Global average sea level rose 0.1–0.2 m, largely due to thermal expansion
- The atmospheric concentrations of carbon dioxide (CO_2) and methane (CH_4) have increased 31% and 151% since 1750
- Precipitation in the Northern Hemisphere (where data are reliable) increased by 1% per decade, and the frequency of heavy precipitation events increased by 2–4% in the century
- Warm episodes of the El Niño–Southern Oscillation phenomenon have been more frequent, persistent and intense since the 1970s compared to the previous 100 years
- Arctic sea-ice thickness during late summer to early autumn has decreased by up to 40%. Global snow cover decreased by about 10% since 1960

Predictions for the 21st century

- Climate models predict that the mean annual global surface temperature will increase by 1.4–5.8 °C over the period 1990–2100, a rate higher than the warming rate experienced in the 20th century
- Warming is likely to be more pronounced over land areas, particularly in high latitudes and in the cold season
- Most models predict a weakening but not a shutdown of the thermohaline circulation by 2100. Shutdown may occur beyond 2100, perhaps irreversibly
- Thermal expansion and loss of ice mass suggest that sea level will rise between 0.09 and 0.88 m by 2100
- The Antarctic ice sheet is likely to gain mass due to greater precipitation, while the Greenland ice mass will decrease because of increased run off

direct as well as indirect effects, and the overall response may often be unexpected, and geographically uneven.

However, if global change research has been conclusive at all it has been to acknowledge that the planet has a tremendous natural variability, and that

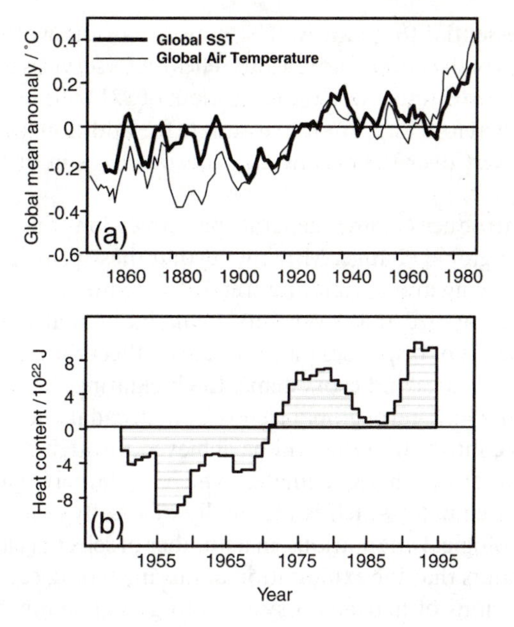

Figure 2 (a) Global air and sea surface temperature variations from 1850 to 1990[6] and (b) heat content of the world's oceans from 1950 to 1995[8]

anthropogenic pressures are an added, if perhaps dominant, force. It is only in the last few decades that we have begun to unravel this natural variability and the multi-decadal, decadal and annual cycles that characterize the Earth systems. To predict the impact of global change on these cycles we must understand their dynamics and linkages, as well as their role in shaping the natural processes of the Earth biota.

This understanding is particularly urgent with reference to the marine systems. The oceans occupy over 70% of the Earth's surface and absorb twice as much of the sun's radiation as the atmosphere or the land surface, playing a major role in shaping the Earth's climate. The oceans move heat from lower to higher latitudes, contribute to the complexity of the Earth's climate and maintain regional differences. However, our knowledge of the functioning of the world's oceans is still limited. Historical information seems to indicate that sea surface temperatures have been warming through the 20th century[6] although at a slower rate than the atmosphere[7] (Figure 2a). Detailed and strong evidence of warming in all the major oceans over the period 1948–1998 has recently been reported[8] (Figure 2b). These results indicate that the mean temperature of the top 300 m of the ocean has increased by 0.31 °C, corresponding to an increase in heat content of approximately 1×10^{23} joules of energy. Furthermore, the warming signal was observable to depths of some 3000 m, demonstrating that the oceans are storing part of the Earth's excess heat.

The biological responses to the warming of the oceans are complex and

[6] M. A. Cane, A. C. Clement, A. Kaplan, Y. Kushnir, D. Pozdnyakov, R. Seager, S. E. Zebiak and R. Murtugudde, *Science*, 1997, **275**, 957–960.

[7] A. Kaplan, M. A. Cane, Y. Kushnir, A. C. Clement, M. B. Blumenthal and B. Rajagopalan, *J. Geophys. Res.*, 1998, **103**, 18 567–18 589.

[8] S. Levitus, J. I. Antonov, T. P. Boyer and C. Stephens, *Science*, 2000, **287**, 2225–2228.

unclear, but essential to quantify. The majority of the world's population lives within 50 miles of the coastal ocean and relies on its services for its support. These services have been estimated to the equivalent of $21 trillion per year[9] (coasts and oceans). It is therefore important to estimate the additional pressures that global change may exert over the oceans and coastal zones in order to preserve their services.

Table 2 introduces some general principles on the expected biological adaptations to global change. Moving beyond these principles requires a major step from observing and correlating trends to identifying cause–effect processes. We have to recognize that every effect on individual species and processes involves a cascade of responses that is likely to affect the structure and dynamics of whole communities and ecosystems. In elucidating these causal processes we need to separate long-term global trends from decadal and multidecadal natural cycles. In this contribution this will be achieved through detailed analysis of the links between atmospheric climate, which is largely globally driven, and oceanographic climate, which is regionally or locally controlled, as well as the subsequent biological adaptations, in a number of selected case studies. A section on the constraints that the exploitation of marine biological resources may exert on the adaptations of marine ecosystems to global change will follow.

The Thermohaline Circulation

Marine communities live in a three-dimensional environment where the depth axis may determine diametrically opposite conditions and control mechanisms. The conventional ocean circulation is characterized by a wind-driven upper circulation that gives rise to massive, near-surface flows such as the Gulf Stream and the Kuroshio and Antarctic Circumpolar currents. Superimposed upon this circulation is the so-called thermohaline circulation (THC). This is driven by surface-ocean density contrasts arising from temperature and salt variations produced by strong atmospheric cooling and wind-induced evaporation. The THC transports huge amounts of heat from the equator to the poles. It has enormous consequences for the weather of Western Europe, warming parts of the continent by up to 10 °C in mean temperature.[11] To balance this transport, water sinks 2–3 km after cooling, and flows back towards the equator along the ocean bottom, eventually returning to the surface, primarily near the ocean boundaries.[12] Through this process oxygen is pumped into the deep sea, sustaining life at depth.

Concerns over the potential consequences of global warming on the THC[13] were initially ignored,[14] but evidence that the THC has closed down in the past has changed this perception.[15–17] The THC is very sensitive to the amount of

[9] R. Costanza, R. d'Arge, R. de Groot, S. Farber, M. Grasso, B. Hannon, K. Limburg, S. Naeem, R. V. O'Neill, J. Paruelo, R. G. Raskin, P, Sutton and M. van den Belt, *Nature*, 1997, **387**, 253–260.

[10] L. Hughes, *Tree*, 2000, **15**, 56–61.

[11] S. Manabe and R. J. Stouffer, *J. Clim.*, 1988, **1**, 841–866.

[12] G. D. Egbert and R. D. Ray, *Nature*, 2000, **405**, 775–778.

[13] W. S. Broecker, *Nature*, 1987, **328**, 123–126.

[14] J. T. Houghton, L. G. Meira Filho, B. A. Callander, N. Harris, A. Kattenberg and K. Maskell, *Climate Change*, Cambridge University Press, 1995.

[15] D. Seidov and M. Maslin, *Geology*, 1999, **27**, 23–26.

Table 2 Generalized predictions of the effects of global environmental change on species and habitats[10]

(*a*) *Effects on physiology* – Influences of temperature, precipitation or/and atmospheric gas composition on metabolic and developmental rates, and processes such as photosynthesis
(b) *Effects on distributions* – Species are expected to move upwards in elevation and towards the poles in latitude in response to global warming
(c) *Effects on phenology* – Life cycle events triggered by environmental cues may be altered, leading to decoupling of phenological relationships between species
(d) *Adaptation* – species with short generation times and rapid population growths might undergo microevolutionary changes

freshwater entering the North Atlantic, but its response is not linear. It has been hypothesized[18] that the North Atlantic has two possible equilibrium points, separated by a threshold point where the circulation breaks down completely. We do not know how close we are at present to this threshold, but common consensus indicates that a decline in turnover rate of 20–50% in the THC circulation is possible by the end of the 21st century[19] (Figure 3). This decline appears to be more sensitive to rapid rather than slow warming.[20] A significant reduction in the deep flow from the Nordic Seas has already been observed,[21] suggesting that the global thermohaline circulation is already weakening. A severe slowdown or shutdown of the THC would increase the rate of sea level rise,[22] and would reduce further the ability of the ocean to take up CO_2,[23] further enhancing global warming.

The Biological Pump

The physical and biological processes that govern the cycling and transport of matter from the surface to the deep sea are commonly referred to as the solubility pump and the biological pump (Figure 4).[24] Both pumps act to increase CO_2 concentrations in the ocean interior. Upon dissolution in water CO_2 forms a weak acid that reacts with carbonate anions and water to form bicarbonate. The capacity of the ocean's bicarbonate system to buffer changes in CO_2 is limited by the addition of Ca^{2+} from the slow weathering of rocks, and the capacity of the ocean to store CO_2 is thus constrained.

The slow overturning of the thermohaline circulation and the seasonal changes in ocean ventilation drive the solubility pump. Cold and dense water masses in high latitude oceans, particularly of the North Atlantic and Southern Ocean, absorb atmospheric CO_2 before sinking to the ocean interior. This sinking is balanced by upwelling in other regions. Upwelled water warms when it reaches

[16] A. Hall and R.J. Stouffer, *Nature*, 2001, **409**, 171–175.
[17] A. Ganopolski and S. Rahmstorf, *Nature*, 2001, **409**, 153–158.
[18] S. Rahmstorf, *Clim. Change*, 2000, **46**, 247–256.
[19] S. Rahmstorf, *Nature*, 1999, **399**, 523–524.
[20] R.J. Stouffer and S. Manabe, *J. Clim.*, 1999, **12**, 2224–2237.
[21] B. Hansen, W.R. Turrell and S. Østerhus, *Nature*, 2001, **411**, 927–930.
[22] R. Knutti and T.F. Stoecker, *J. Clim.*, 2000, **13**, 1997–2001.
[23] J.L. Sarmiento and C. Le Quéré, *Science*, 1996, **274**, 1346–1350.
[24] D. Baird and R. Ulanowicz, *Ecol. Monogr.*, 1989, **59**, 329–364.

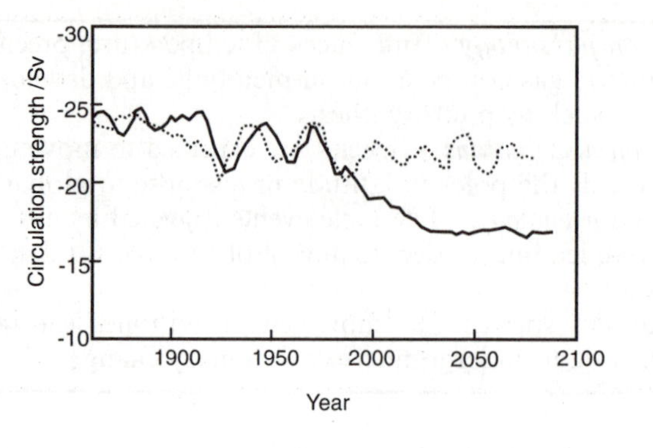

Figure 3 Results of model HadCM3 of the Hadley Centre for Climate Prediction on the strength of the North Atlantic Ocean circulation. Dotted line, control simulation. Solid line: results of a non-interventionist greenhouse gas emissions scenario. (From http://www.metoffice.com/research/hadleycentre/index.html)

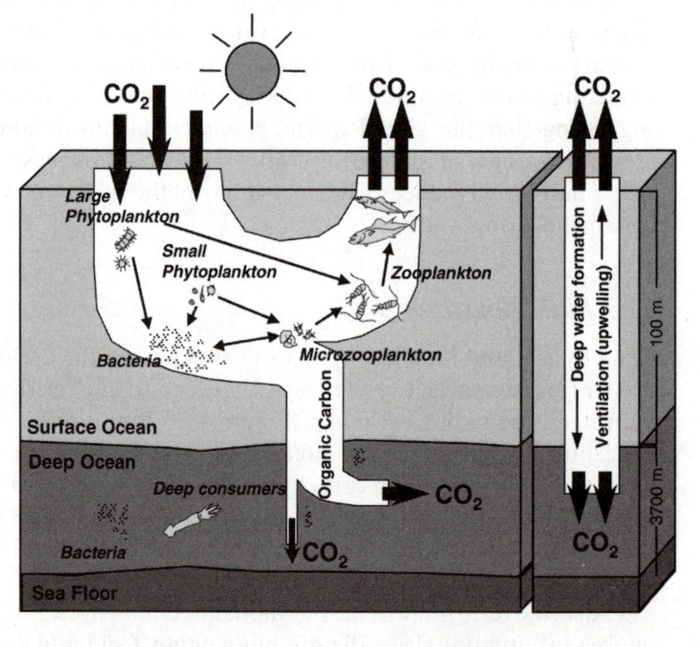

Figure 4 Schematic view of the biological pump (left box) and the solubility pump (right box)[24]

the surface, and as a result CO_2 becomes less soluble and some is released back to the atmosphere.

The biological pump plays a major role in limiting the atmospheric concentrations of CO_2, through photosynthetic processes, but the transfers of carbon inside the biological pump are still poorly quantified, largely because of the complexity and uniqueness of most marine food webs. For example, no less than 36 different carbon compartments characterize the Chesapeake Bay ecosystem.[25] Quantification of the standing stock and transfer rates for each and every compartment is necessary before the direct and indirect consequences of an external physical

[25] S. W. Chilsholm, *Nature*, 2000, **407**, 685–687.

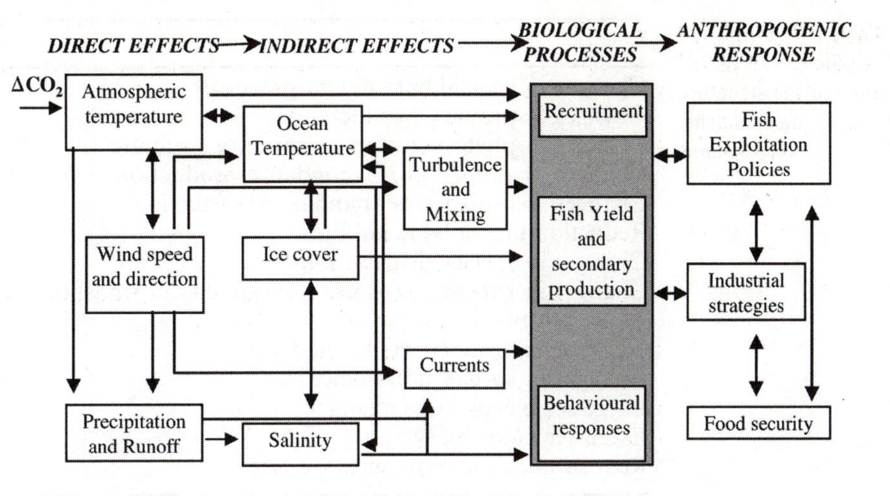

Figure 5 Major climatic pathways linking an increase in atmospheric CO_2 to the abiotic and biotic environment of fishes, and their anthropogenic consequences. The shaded area identifies the core objective of GLOBEC

forcing, such as increasing atmospheric CO_2 concentrations or surface temperatures, can be understood.

2 Climate Variability and Change: Coupling Atmospheric and Biological Processes in the Ocean

The major pathways linking atmospheric processes to biological changes in the marine environment are identified in Figure 5, which is the basis for the development of GLOBEC (Global Ocean Ecosystem Dynamics), a major international research programme aimed at understanding the structure and dynamics of the marine ecosystem in the context of climate change (www.globec.org). Changes in water temperature, precipitation and wind would generally affect water stratification and mixing processes. These, in turn, would have implications for the retention of organisms, regeneration of nutrients and production processes. For example, changes in wind-driven vertical mixing would determine the mixed layer depth and, as a result, the primary production in it. The effects are expected to cascade through higher trophic levels, from secondary producers to humans.

There is significant literature to support the contention that global change is influencing biological processes in the ocean at all scales (Table 3). However, many of these examples are inconclusive because the processes behind the observations are not fully understood. It may well be that the most prominent climate impacts on marine ecosystems, as we have experienced them up to the present time, are not caused by the long-term anthropogenically generated climate changes, but by those phenomena linked to ocean climate fluctuations of interannual to decadal scales. It is therefore essential to understand how these fluctuations affect biological processes in the sea, while at the same time to ascertain how global change will influence these fluctuations.

It is understood that the climate of many areas of the world is regulated by a number of semi-permanent pressure Centres of Action (CoAs). In the Northern hemisphere these are the Aleutian Low, the Asiatic High, the Pacific High, the

Table 3 Examples of consequences of global change in the structure and functioning of marine ecosystems

Effects	*References*
Changes in coastal enrichment processes	26, 27
Changes in primary productivity	28, 29
Changes in phytoplankton community structure	30, 31
Changes in zooplankton abundance/production	32–34
Changes in fish species abundance/production	31, 35, 36
Reduction in available habitats	37
Changes in species distributions	33, 38–44
Changes in pathways/intensity/seasonality of transport of prey to predators	45–47
Negative cascade effects to predators	48–52
Changes in species dominance	31, 35, 53–55
Large-scale ecosystem changes	56–59
Extensive coral bleaching	60
Reduction in coral calcification rate	61, 62

Iceland Low and the Azores High. Similar CoAs are identified for the southern hemisphere. Based on these CoAs a number of pressure indices have been computed and are used regularly to explain weather and ocean climate patterns at the basin scale. In this review I will concentrate on three of these indices, in order to exemplify how they affect biological production in the sea, and in turn how they are affected by global change. These selected indices are the North Atlantic Oscillation, NAO, the El Niño-Southern Oscillation, ENSO, and the Pacific Decadal Oscillation, PDO. This is not an arbitrary selection. The NAO has a period of approximately 10 years, and is responsible for a number of climatic and biological processes in the North Atlantic. Presently climate models do not predict any discernible trend in the NAO index associated with global warming. On the other hand the ENSO has a periodicity of months to years, and its frequency may be increasing as a result of global warming.[63] Its effects are felt

[26] R. R. Dickson, P. M. Kelly, J. M. Colebrook, W. S. Wooster and D. H. Cushing, *J. Plankton Res.*, 1988, **10**, 151–169.

[27] A. Bakun, *Science*, 1990, **247**, 198–201.

[28] P. C. Reid, M. Edwards, H. G. Hunt and A. Warner, *Nature*, 1998, **391**, 546.

[29] Bering Sea Task Force, *Status of Alaska's Oceans and Marine Resources*, Bering Sea Task Force Report to Governor Tony Knowles, 1999.

[30] T. C. Vance, J. D. Schumacher, P. J. Stabeno, C. T. Baier, T. Wyllie-Echeverria, C. T. Tynan, R. D. Brodeur, J. M. Napp, K. O. Coyle, M. B. Decker, G. L. Hunt Jr, D. Stockwell, T. E. Whitledge, M. Jump and S. Zeeman, *EOS*, 1998, **79**, 121–126.

[31] G. H. Kruse, *Alaska Fishery Res. Bull.*, 1998, **5**, 55–63.

[32] A. Conversi, S. Piontkovski and S. Hameed, *Deep-Sea Res. II*, 2001, **48**, 519–530.

[33] S. J. Holbrook, R. J. Schmitt and J. S. Stephens Jr, *Ecol. Appl.*, 1997, **7**, 1299–1310.

[34] D. Roemmich and J. McGowan, *Science*, 1995, **267**, 1324–1326.

[35] R. D. Brodeur, P. A. Kruse, P. A. Livingstone, G. Walters, J. Ianelli, G. L. Swartzman, M. Stepanenko and T. Wyllie-Echeverria, in *Report of the FOCI International Workshop on Recent Conditions in the Bering Sea*, 1999, ed. S. A. Macklin, *Contribution 2044 from NOAA/Pacific Marine Environmental Laboratory, Contribution B358 from Fisheries-Oceanography Coordinated Investigations.*

[36] D. M. Ware and G. A. McFarlane, in *Clim. Change and North. Fish Populations. Can. Spec. Publ. Fish. Aquat. Sci.*, ed. R. J. Beamish, 1995, **121**, 509–521.

[37] D. W. Welch, Y. Ishida and K. Nagasawa, *Can. J. Fish. Aquat. Sci.*, 1998, **55**, 937–948.

throughout the world, but mainly in the Equatorial Pacific. Finally, the PDO has cycles of 20–30 years, and is responsible for processes in the North and Central Pacific. Because of the length of its cycles we are not in a position to predict whether global change is affecting its periodicity.

Case Study 1: The North Atlantic Oscillation

The NAO is a basin-wide scale atmospheric alternation of atmospheric mass, reflecting the changes in sea level pressure over the North Atlantic region.[64] The standard NAO index (Figure 6) is the difference in normalized sea level pressures between the subtropical high pressures centred on the Azores and the sub-polar low pressures centred over Iceland, during the winter season (December to March). Its state determines the speed and direction of the westerlies across the north Atlantic as well as the level and direction of moisture transport and winter temperatures on both sides of the basin.[64] A high NAO index is associated with

[38] J. P. Barry, C. H. Baxter, R. D. Sagarin and S. E. Gilman, *Science*, 1995, **267**, 672–675.

[39] A. M. Breeman, in J. J. Beukema and J. J. W. M. Brouns (eds.) *Expected Effects of Climatic Change on Marine Coastal Ecosystems*, 1990, Kluwer Academic Press, pp. 69–76.

[40] K. T. Frank, I. Perry and K. F. Drinkwater, *Trans. Am. Fish. Soc.*, 1990, **119**, 353–365.

[41] J. Lubchenco, S. A. Navarrete, B. N. Tissot and J. C. Castilla, in H. A. Mooney, E. R. Fuentes and B. I. Kronberg, *Earth System Responses to Global Change*, 1993, pp. 147–166.

[42] S. A. Murawski, *Trans. Am. Fish. Soc.*, 1993, **122**, 647–658.

[43] A. J. Southward, *J. Mar. Biol. Assoc. UK*, 1967, **47**, 81–95.

[44] A. J. Southward, S. J. Hawkins and M. T. Burrows, *J. Therm. Biol.*, 1995, **20**, 127–155.

[45] D. H. Cushing, *Climate and Fisheries*, Academic Press, London, 1982, 387 pp.

[46] M. R. Heath, J. O. Backhaus, K. Richardson, E. Mckenzie, D. Slagstad, D. Beare, J. Dunn, J. G. Fraser, A. Gallego, D. Hainbucher, S. Hay, S. Jonasdottir, H. Madden, J. Mardaljevic and A. Schacht, *Fish. Oceanogr.*, 1999, **8**(Supp. 1), 163–176.

[47] B. Planque and A. H. Taylor, *ICES J. Mar. Sci.*, 1998, **55**, 644–654.

[48] D. G. Ainley, W. J. Sydeman and J. Norton, *Mar. Ecol. Prog. Ser.*, 1996, **118**, 69–79.

[49] K. J. Kuletz, D. B. Irons, B. A. Angler and J. F. Piatt, *Proc. Int. Symp. Role of Forage Fishes in Mar. Ecosys.*, 1997, 703–706.

[50] J. M. Napp, K. O. Coyle, T. E. Whitledge, D. E. Varela, M. V. Flint, N. Shiga, D. M. Schell and S. M. Henrichs, in *Report of the FOCI International Workshop on Recent Conditions in the Bering Sea*, ed. S. A. Macklin, 1999, *Contribution 2044 from NOAA/Pacific Marine Environmental Laboratory, Contribution B358 from Fisheries-Oceanography Coordinated Investigations.*

[51] I. Stirling, N. J. Lunn and J. Iacozza, *Arctic*, 1999, **52**, 294–306.

[52] R. R. Veit, P. Pyle and J. A. McGowan, *Mar. Ecol. Prog. Ser.*, 1996, **139**, 11–18.

[53] J. P. Croxall, K. Reid and P. A. Prince, *Mar. Ecol. Progr. Ser.*, 1999, **177**, 115–131.

[54] W. R. Fraser, W. Z. Trivelpiece, D. G. Ainley and S. G. Trivelpiece, *Pol. Biol.*, 1992, **11**, 525–531.

[55] D. Lluch-Belda, S. Hernandez-Vazquez, D. B. Lluch-Cota, C. A. Salinas-Zavala and R. A. Schwartzlose, *CalCOFI Rep.*, 1992, **33**, 50–59.

[56] N. P. Holliday and P. C. Reid, *ICES J. Mar. Sci.*, 2001, **58**, 270–274.

[57] W. A. Montevecchi and R. A. Myers, *ICES J. Mar. Sci.*, 1997, **54**, 608–614.

[58] P. C. Reid, M. de Fatima Borges and E. Svendsen, *Fish. Res.*, 2001, **50**, 163–171.

[59] R. D. Sagarin, J. P. Barry, S. E. Gilman and C. H. Baxter, *Ecol. Monographs*, 1999, **69**, 465–490.

[60] C. Wilkinson, O. Linden, H. Cesar, G. Hodgson, J. Rubens and A. E. Strong, *Ambio*, 1999, **28**, 188–196.

[61] J. P. Gattuso, D. Allemand and M. Frankignoulle, *Am. Zool.*, 1999, **39**, 160–183.

[62] J. A. Kleypas, R. W. Buddemeir, D. Archer, J. P. Gattuso, C. Langdon and B. N. Opdyke, *Science*, 1999, **284**, 118–120.

[63] A. Timmermann, J. Oberhuber, A. Bacher, M. Esch, M. Latif and E. Roeckner, *Nature*, 1999, **398**, 694–696.

[64] J. W. Hurrell, *Science*, 1995, **269**, 676–679.

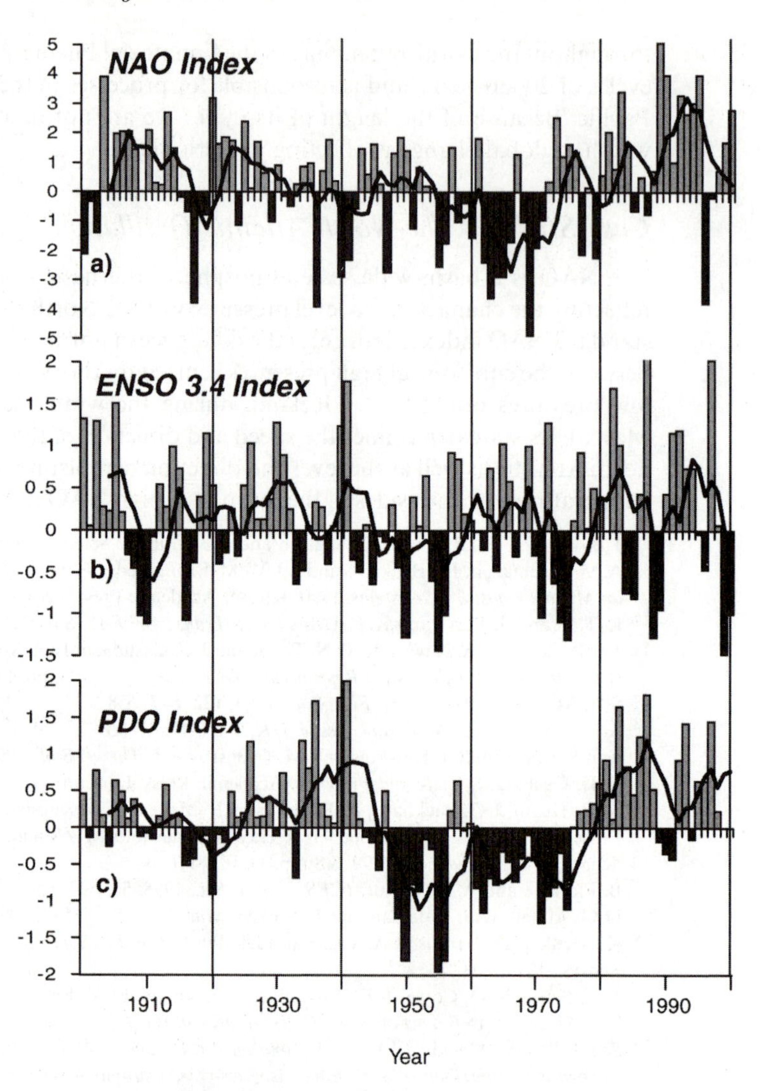

Figure 6 (a) Winter (December to March) index of the NAO between Lisbon and Reykjavik.[64] (b) Annual mean El Niño Southern Oscillation 3.4 index, based on SST anomalies in the region 120 °W–170 °W, 5 °S–5 °N.[66] (c) PDO index, derived as the leading PC of monthly SST anomalies in the North Pacific Ocean, poleward of 20 °N.[91] (a) and (b) courtesy of the National Centre for Atmospheric Research (NCAR), http://www.cgd.ucar.edu/cas/climind/. (c) Courtesy of Steven Hare, Pacific Halibut Commission, http://www.iphc.washington.edu/staff/hare/html/decadal/post1977/pdo1.html

stronger than normal wind circulation in the north Atlantic, high temperatures in Western Europe and low temperatures on the East Coast of Canada. Recent simulations indicate that the NAO index can be reconstructed from knowledge of North Atlantic Sea Surface temperature records, identifying a major feedback loop between ocean and atmosphere.[65]

The North Atlantic is a convective ocean, producing NADW (North Atlantic Deep Water) in the Labrador Sea and Greenland/Iceland Sea, and driving the global thermohaline circulation. Ice-core data have revealed large-scale decadal climate variability in the North Atlantic, which can be related to the North Atlantic Oscillation, NAO.[67] The NAO is therefore responsible for modulating

[65] M.J. Rodwell, D.P. Rowell and C.K. Folland, *Nature*, 1999, **398**, 320–323.
[66] K.E. Trenberth, *Bull. Am. Meteor. Soc.*, 1997, **78**, 2771–2777.
[67] L.K. Barlow, J.W.C. White, R.G. Barry, J.C. Rogers and P.M. Grootes, *Geophys. Res. Newsl.*, 1993, **20**, 2901–2904.

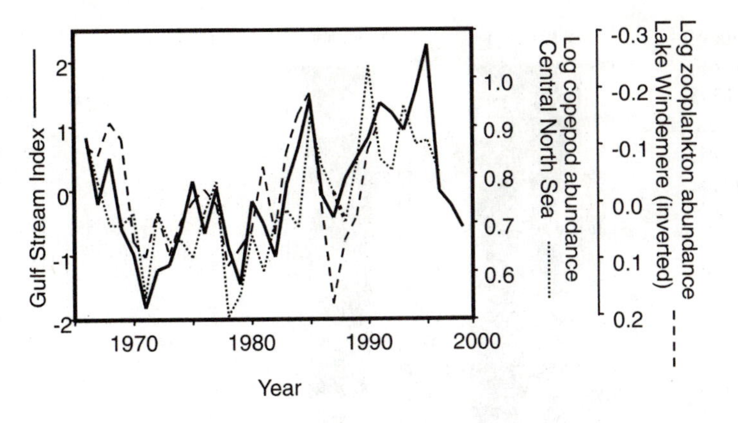

Figure 7 Annual values of the Gulf Stream first principal component (arbitrary units) compared with the number of copepods in the Central North Sea from the Continuous Plankton Recorder Surveys[69] and mean summer biomass of zooplankton in the North basin of Lake Windemere[71]

and coordinating convective activity in the North Atlantic.[68]

The NAO has a discernible influence on the dynamics of many physical and biological processes, both in the ocean and on land. The Gulf Stream, part of the global ocean conveyor belt, is largely responsible for the distribution of heat in the Northern Hemisphere. High NAO years result in the Gulf Stream following a more northerly path 2 years later, indicating a direct influence of atmospheric climate on the ocean climate.[47] The northerly extent of the Gulf Stream has in turn been correlated with the abundance of zooplankton in the UK and surrounding areas,[69] crop yields and productivity of natural vegetation[70] as well as zooplankton production in UK lakes[71] (Figure 7). The timing of the spring phytoplankton bloom in the North Atlantic,[72] as well as in central European lakes,[73,74] appears linked to the NAO. This response is not homogeneous though, and varies from lake to lake, depending on their thermal structure and mixing regime.[75] These examples highlight the overall influence of the NAO on individual biological processes, an influence that extends over the functioning of the entire North Atlantic ecosystem.

A remarkable parallelism between long-term trends in North Atlantic westerly winds and four marine trophic levels (phytoplankton, zooplankton, fish and marine birds) has been observed.[76] The authors concluded that the mechanisms that produce such trends were likely to be considerably more complicated than resulting from trophic interactions carried through the food chain only. More recently, dramatic biological changes in the ecology of the North Sea and the Central East Atlantic around 1988 were observed, coinciding with the highest positive NAO index records for more than a century.[28] It was observed that phytoplankton abundance and the frequency of blooms had increased drastically

[68] R. Dickson, J. Lazier, M. Meincke, P. Rhines and J. Swift, *Prog. Oceanog*, 1996, **38**, 241–295.
[69] A.H. Taylor, *ICES J. Mar. Sci.*, 1995, **52**, 711–721.
[70] A.J. Willis, N.P. Dunnett and J.P. Grime, *OIKOS*, 1995, **73**, 408–410.
[71] D.G. George and A.H. Taylor, *Nature*, 1995, **378**, 139.
[72] X. Irigoien, R.P. Harris, R.N. Head and D. Harbour, *J. Plank. Res.*, 2000, **22**, 2367–2371.
[73] D. Gerten and R. Adrian, *Limnol. Oceanogr.*, 2000, **45**, 1058–1066.
[74] G. Weyhenmeyer, T. Blenckner and K. Pettersson, *Limnol. Oceanogr.*, 1999, **44**, 1788–1792.
[75] D. Gerten and R. Adrian, *Limnol. Oceanogr.*, 2001, **46**, 448–155.
[76] N.J. Aebischer, J.C. Coulson and J.M. Colebrook, *Nature*, 1990, **347**, 753–755.

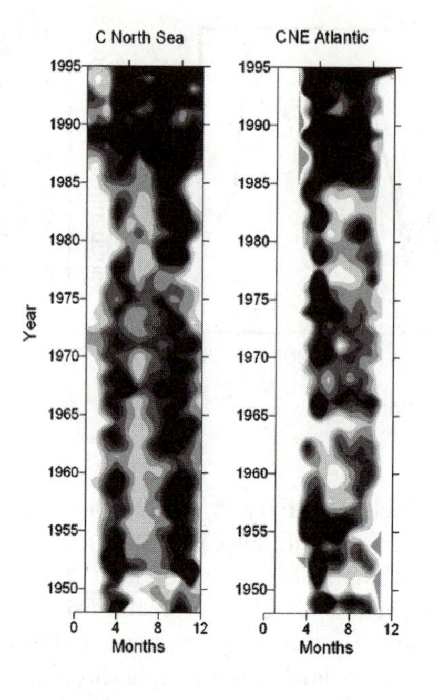

Figure 8 Contour plots of monthly phytoplankton colour during 1948–1995 for the Central North Sea, central northeast Atlantic and northern northeast Atlantic.[28]

(Figure 8). Zooplankton abundance also increased (lagged 1 year), as did horse mackerel catches[58] (lagged two years) (Figure 9). It is now believed that these are indicators of a major regime shift that affected the North Atlantic around 1988. Alterations in the centre of deep water convection from the Greenland Sea to the Labrador Sea after 1988,[68] and increases in the flow of oceanic water into the North Sea through the Shelf Edge Current[56] appear to be the main drivers for these changes. A similar shift may have also been observed after 1996, extending further north the habitat of tropical fish species west of Europe.[77] It is unclear whether these shifts are indicative of long-term global change or short-term stochastic variability, but indicate the potential for specific atmospheric physical forcings to dramatically influence the structure and productivity of the North Atlantic ecosystem.

Non-linear responses to physical forcing have also been observed in this region. For example, the NAO index and the abundance of the copepod *Calanus finmarchicus* in the North East Atlantic were negatively correlated from 1958 to 1995,[78] although positively correlated in the Northwest Atlantic.[32] When the NAO reversed to negative values in 1996–97, the abundance of *C. finmarchicus* in the Northeast Atlantic remained low[79] (Figure 10) and is yet to recover. It is now believed that the reversal of the NAO did not result in any immediate increase in the supply of Norwegian Sea Deep Water (NSDW) to the Faroe–Shetland channel bottom water. In consequence, the supply of copepods has remained at a low level, despite the more favourable climatic conditions. It has been postulated that it would take several years of persistent deep convection in the Greenland

[77] J.-C. Quero, M. H. Du Buit and J. J. Vayne, *Oceanol. Acta*, 1998, **21**, 345–351.
[78] J. M. Fromentin and B. Planque, *Mar. Ecol. Prog. Ser.*, 1996, **134**, 111–118.
[79] B. Planque and P. C. Reid, *J. Mar. Biol. Assoc. UK*, 1998, **78**, 1015–1018.

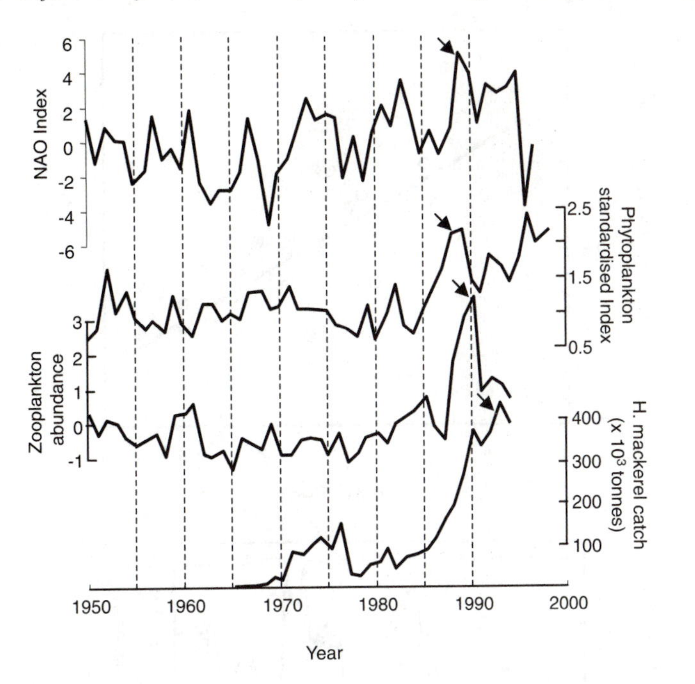

Figure 9 Time series of North Sea indicators: NAO index, phytoplankton colour index, principal component of zooplankton abundance from the CPR surveys[56] and horse mackerel for the total area 45–65 °N[58]

Sea to restore the North Atlantic Deep Water (NADW) overflow and produce a recovery in *Calanus finmarchicus*.[46] Such results are invaluable in understanding the magnitude and direction of potential changes from global environmental change.

The relationship between *C. finmarchicus* and deep-water transport in the North Atlantic has consequences for higher trophic levels, and in particular for cod, *Gadhus morhua*. Cod is the main exploited fish species in the North Atlantic and the most intensively studied fish species worldwide. On the one hand global warming generally benefits the recruitment of northern cod stocks and adversely affects recruitment of southern stocks, because cod recruitment is optimal at 8 °C.[80] But as young cod's food supply depends on the availability of *C. finmarchicus* its responses to global change may be more complex. It has been postulated that Atlantic cod stocks are generally distributed around the rim of the two major habitats of *C. finmarchicus*, within the North Atlantic subpolar gyre.[81] The northern stocks of cod would have a lower ambient temperature than at the core *Calanus* region, and the southern cod stocks would have a higher temperature than at the core region. Advection from the copepod-rich region to the edges would bring with it colder water to the Southern and warmer water to the Northern stocks, bringing water temperatures closer to optimal while at the same time improving the prey field for cod. Whether cod is more affected by the more suitable temperature field or by the enhanced prey field may appear academic, but in fact it would pinpoint the actual process that translates physical forcing into biological energy. The compounded effect of fishing pressure on these processes adds a new dimension, and justifies the importance that GLOBEC places in incorporating anthropogenic influences in the study of biological

[80] B. Planque and T. Fredou, *Can. J. Fish. Aquat. Sci.*, 1999, **56**, 2069–2077.
[81] S. Sundby, *Sarsia*, 2000, **85**, 277–298.

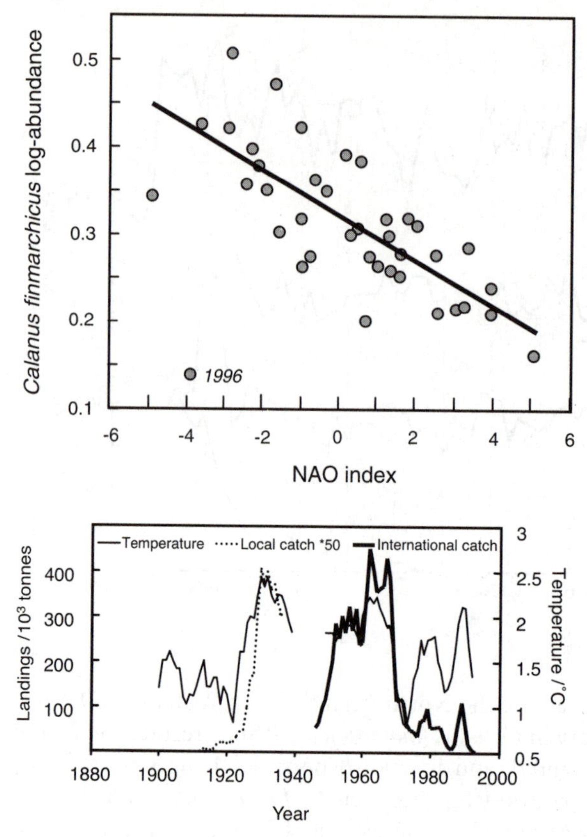

Figure 10 Relationship between annual abundance of *Calanus finmarchicus* in the Northeast Atlantic and the NAO index[78]

Figure 11 Cod catches and water temperature at West Greenland[83]

adaptations to global change. Total cod catches have increased steadily through the 20th century, to peak at 3.9 million tonnes in 1968 and declining ever since (Figure 11 and ref. 83). It is now believed that the combination of an unfavourable environment and severe fishing pressure may have reduced the population of North Sea cod to a small number of young, immature fish, extremely limited in their ability to respond to favourable environments.[82]

Case Study 2: The El Niño-Southern Oscillation

Our second case study will focus on the El Niño off the equatorial Pacific. The ocean circulation in this tropical region is governed by the trade winds, which blow intensely along the equator towards the Australian–Indonesian low-pressure zone. As a result warm air rises in that zone and returns eastwards, sinking over the cold dry South Pacific high-pressure system (Figure 12). The trade winds drive the westward-flowing South Equatorial Current (SEC), thus producing high sea levels and a large pool of warm water in the western Pacific basin. This is compensated for by the development of coastal upwelling, bringing cold, nutrient-rich water to the coastal areas of the eastern Pacific. In certain years,

[82] C.M. O'Brien, C.J. Fox, B. Blanque and J. Casey, *Nature*, 2000, **404**, 142.

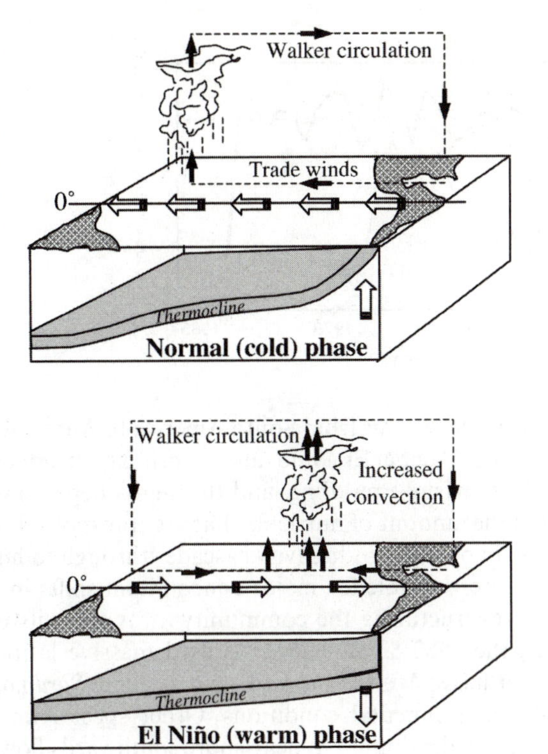

Figure 12 Schematic representation of the atmospheric and ocean processes along the equatorial Pacific during normal and El Niño conditions

around November/December, the trade winds weaken, causing a severe disturbance in the ocean–atmosphere balance. As a result the warm pool of water in the western Pacific shifts to the central and eastern parts of the basin, and the coastal upwelling that dominates this region fails to develop (Figure 12). This phenomenon is called El Niño, and occurs with an average periodicity of 2–7 years.

The term ENSO (El Niño Southern Oscillation) was coined to reflect the close relationship between Pacific El Niño warming events and warming episodes of the so-called Southern Oscillation Index (SOI). The Southern Oscillation is the oscillatory swaying of pressure backwards and forwards between the Indian and Pacific Oceans.[84] The sea level pressure anomalies in the Australian–Indonesian low-pressure zone tend to be of opposite sign to those in the South Pacific high-pressure zone. The SOI is the pressure difference between these two regions. There is a nearly one-to-one correspondence between the warm episodes of the SOI and the South American El Niño, with the latter leading the former by 4–8 months.[85]

The effects of El Niño are vast and dramatic. In terrestrial ecosystems the changes in precipitation cause the greening of deserts and the crash of agricultural crops. Vegetation changes then create a cascade of effects that affect

[83] K. Brander, *Soc. Exp. Biol. Ser.*, 1996, **61**, 255–278.
[84] P. B. Wright, *Bull. Am. Meteorol.*, 1985, **66**, 398–412.
[85] J. D. Horel and J. M. Wallace, *Mon. Weather Rev.*, 1981, **109**, 813–829.

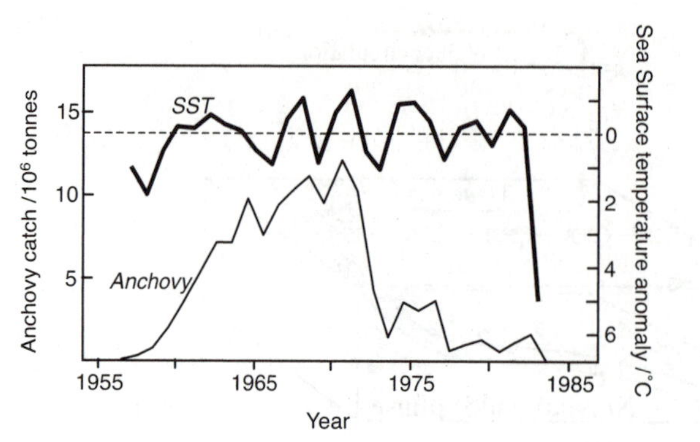

Figure 13 Relationship between the annual anomaly of Sea Surface Temperature (SST) and the annual catch of anchovy off Peru. The anomaly scale is inverted: downwards indicates warmer (El Niño) phases[87]

whole terrestrial ecosystems from South America to Africa (see review in ref. 86). In the equatorial Pacific El Niño affects primary productivity through the reduction of the coastal upwelling, and through a depression of the nutricline, which reduces the amount of nutrients that are transported to surface waters.[87] This reduction in overall productivity cascades through to higher trophic levels. El Niño is also responsible for major concomitant shifts in the distribution of some species, re-structuring the community away from its equilibrium points. For example, the 1982/83 El Niño[87] caused massive latitudinal shifts in the distributions of hake, *Merluccius gayi*, and sardine, *Sardinops sagax*, to avoid unfavourable environmental conditions. Other species were simply removed from the system by the dominant near-shore southwards flow that characterizes El Niño, such as the shrimps *Xiphopenaeus riueti* and *Penaeus occidentalis*. A combination of habitat reduction and increased predation by species favoured by El Niño conditions (like bonito, *Sarda chiliensis*, dorado, *Coryphaena hippurus*, and yelowfin, *Thunnus albacores*) are thought to have affected the abundance of other major species, like jack mackerel, *Trachurus symmetricus*. But the most affected marine species was probably the anchovy, *Engraulis ringens*, the world's largest fishery (Figure 13). The 20-fold reduction in primary production severely limited anchovy growth and survival. Other components of the ecosystem showed positive responses, such as the scallop, *Argopecten purpuratus*, possibly because of the metabolic advantages of warmer conditions.

The ENSO signature can be found in oceanographic processes beyond the tropical Pacific as well. For example the path and intensity of the Gulf Stream is generally forecastable from the intensity of the North Atlantic Oscillation Index (NAO), which accounts for half of its variance.[88] Much of the remaining variance is accounted for by the Southern Oscillation in the Pacific. The Gulf Stream appears to be displaced northwards following ENSO events,[89] and the total freshwater export from the Atlantic is about 0.1 Sverdrups (Sv) larger during El Niño events.[90] Therefore changes in the South Pacific can, through global teleconnections, influence oceanographic conditions in the North Atlantic,

[86] M. Holmgren, M. Scheffer, E. Ezcurra, J. R. Gutierrez and G. M. J. Mohren, *Tree*, 2001, **16**, 89–94.
[87] R. T. Barber and F. P. Chavez, *Nature*, 1986, **319**, 279–285.
[88] A. H. Taylor and J. H. Stephens, *Tellus*, 1998, **50A**, 134–142.
[89] A. H. Taylor, M. B. Jordan, J. A. Stephens, *Nature*, 1998, **393**, 638.
[90] A. Schmittner, C. Appenzeller and T. F. Stocker, 2000, *Geophys. Lett.*, **27**, 1163–1166.

supporting the need to conduct ecosystem comparisons to understand the causes of observed changes in particular ecosystems.

Concerns have recently been raised that global warming is likely to affect El Niño events by altering the background climate,[91] as ENSO involves large redistribution of heat in the tropical Pacific. Models suggest that increased greenhouse gases are likely to increase the frequency of El Niños as well as the intensity of cold periods,[63] but the level of greenhouse gases that will cause such changes is debated.[92] From past records we know that ENSOs have existed for at least 130 000 years, but their strength seems to have increased in the 20th century.[93] At shorter time scales it has been noted that in the cooler and drier tropical ocean of the late 19th century ENSO cycles lasted about 10–15 years, which were replaced by strong shorter cycles (3 years) coinciding with the warming step in the early 20th century.[94] Although three decades of weak interannual variability followed until 1950, the results suggest that ENSO may respond to further global warming in ways that we still do not understand and that may be more complex than we anticipate. Recent results suggest that ENSO events can also interact with changes in the Earth's orbit to trigger rapid changes in climate.[95] Periodic and gradual changes in the shape of the Earth's path around the Sun and the tilt of its axis are thought to cause radical climate shifts such as the ice ages. These orbital variations can alter ENSO's pulse, locking it into step with the yearly cycle of seasons. During such a deadlock average temperatures across the globe fall significantly. The authors believe that this changeover between normal (2–7 year periodicity resulting from heat and energy moving between atmosphere and oceans) to locked ENSO happens in a matter of decades. The current ENSO cycle seems to be in a phase close to that in which sudden switches can occur. If correct, ENSO and climate change may be closer that we currently perceive them to be.

Case Study: The Pacific Decadal Oscillation

The Pacific Decadal Oscillation describes a decadal pattern of climate variability in the Pacific. It has similar climatic fingerprints to El Niño, but a significantly different temporal behaviour (Figure 6). PDO eras persist for 20–30 years, while ENSO events are typically 6–18 months long.[96] Warm phases of the PDO are characterized by anomalously cool temperatures in the central North Pacific, and unusually warm temperatures along the west coast of the Americas. At the same time these conditions favour low pressures over the North Pacific and high over western North America and the subtropical Pacific, enhancing counter-clockwise wind stress over the North Pacific.[97] Only two full cycles have been

[91] A. V. Fedorov and S. G. Philander, *Science*, 2000, **288**, 1997–2002.

[92] M. Collins, *J. Clim.*, 2000, **13**, 1299–1312.

[93] A. W. Tudhope, C. P. Chilcott, M. T. McCulloch, E. R. Cook, J. Chapell, R. M. Ellam, D. W. Lea, J. M. Lough and G. B. Shimmield, *Science*, 2001, **291**, 1511–1517.

[94] F. E. Urban, J. E. Cole and J. T. Overpeck, *Nature*, 2000, **407**, 989–993.

[95] A. C. Clement, M. A. Cane and R. Seager. *J. Clim.*, 2001, **14**, 2369–2375.

[96] S. R. Hare, N. J. Mantua and R. C. Francis, *Fish. Habitat*, 1999, **24**, 6–14.

[97] N. J. Mantua, S. R. Hare, Y. Zhang, J. M. Wallace and R. C. Francis, *Bull. Am. Meteorol. Soc.*, 1997, **78**, 1069–1079.

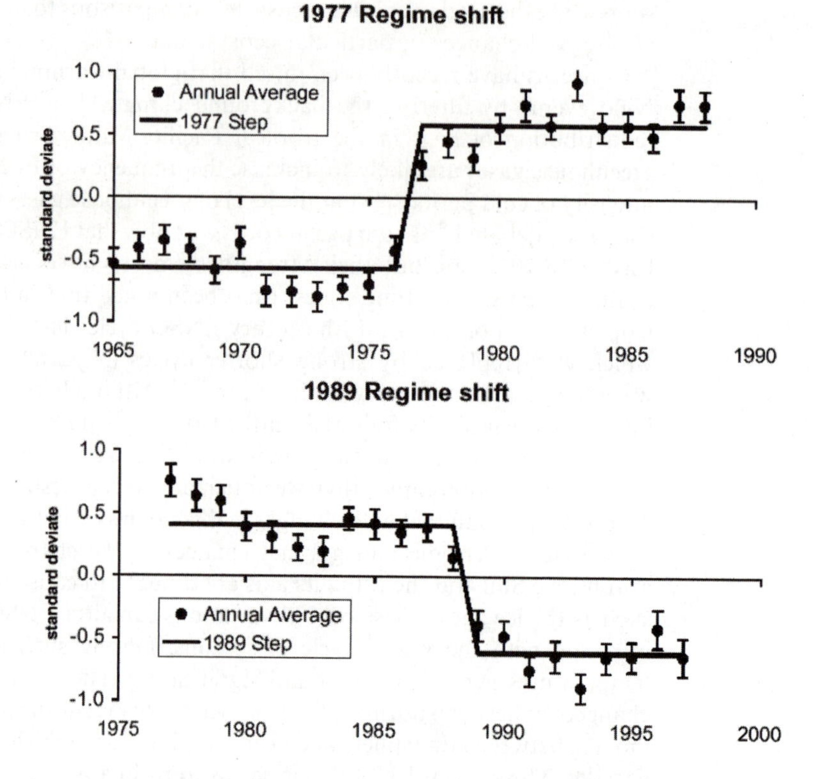

Figure 14 Results from two regime shift analyses of a composite of the 100 environmental time series in the North Pacific. The step passes through the mean standard deviate within each regime. The standard error of the 100 time series is illustrated for each year[98]

observed in the last century: 1890–1924 (cool), 1925–1946 (warm), 1947–1976 (warm) and 1976–at least mid-1990s (cool).[98]

PDO cycles unleash a number of oceanic processes that cause dramatic regime shifts in the productivity of the North Pacific.[96,98–100] Using Principal Component Analysis to identify temporally coherent changes in 100 physical and biological time series of the North Pacific and Bering Sea, two regime shifts in 1977 and 1989 were identified[98] (Figures 14 and 15). They concluded that independent analyses of climate-only and biology-only data matrices point to the 1977 changes being pervasive throughout the Pacific climate and marine ecosystems. The shift included an intensification of the wintertime Aleutian Low, a year-round cooling of the Central North Pacific Ocean, and a year-round warming of the coastal Northeast Pacific Ocean and Bering Sea (Figure 15). Exemplary consequences are the decrease in zooplankton abundance off California,[34] decreases in Alaskan shrimp populations and most west coast salmon populations, and increases in most Alaskan salmon populations.[98] A second shift was observed in 1989, particularly based on biological data. Climatically there was a winter cooling of the coastal waters in the northern Gulf of Alaska and the Bering Sea, a winter warming of the Central North Pacific Ocean, and intensification of the winter

[98] S. R. Hare and N. J. Mantua, *Prog. Oceanogr.*, 2000, **47**, 103–145.

[99] R. J. Beamish, D. J. Noakes, G. A. MacFarlane, L. Klyashtorin, V. V. Ivanov and V. Kurashov, *Can. J. Fish. Aquat. Sci.*, 1999, **56**, 516–26.

[100] T. L. Hayward, *Tree*, 1997, **12**, 150–154.

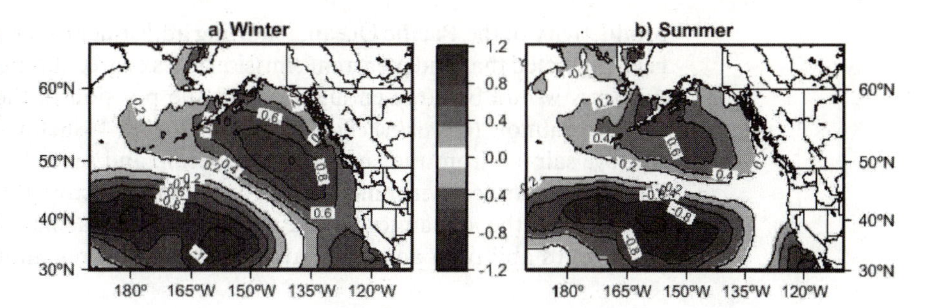

Figure 15 Difference maps for SST change across a possible regime shift in the North Pacific in 1977[98]

and summer Arctic vortex, a weakened Aleutian Low, and a summer warming throughout much of the Central North Pacific and coastal Northeast Pacific Ocean. Important coherent changes included declines in Bering Sea groundfish recruitment, Western Alaska Chinook, chum and pink salmon catch, British Columbia coho, pink and sockeye salmon catch, West coast salmon catches and groundfish recruitment, and increases in Bering Sea jellyfish biomass.[98]

Most importantly, the 1989 change was not a simple reversal of ecosystem conditions established after 1977. While the 1977 shift produced a near-equal balance between fish stocks showing increases and decreases in abundance, the 1989 shift was expressed largely in reductions in productivity. It has been proposed that two energetic inter-decadal climate oscillations, one at a period of 50–70 years, the other at a period of 15–25 years, have been operating in the Pacific Ocean in the 20th century.[101] It was demonstrated that these two oscillations became superimposed on several occasions during the last century, causing major and minor regime shifts consistent with the 1977 and 1989 regime shifts discussed above.

The relevance of the PDO to understanding climate variability is that it shows that 'normal' climate conditions can alter over time periods comparable to a human lifetime. The impact of these regimes shifts is such that salmon runs are now managed and optimal catch levels computed under the assumption that data collected prior to the mid-1970s are no longer relevant to modelling the dynamics of the present-day salmon runs.[96]

The effect of the 1977 regime shift was so abrupt over the entire Pacific Ocean that it opened questions as to whether it was purely a result of natural cycles or whether some component was due to global environmental change. In addition to the effects mentioned earlier, zooplankton biomass also decreased in the Kuroshio–Oyashio Current system[102] and off Peru.[103] Interestingly, anchovy biomass off Peru also decreased, but it was replaced by an increased sardine biomass, indicating how similar species within an ecosystem can respond quite differently to climate forcing. If the shift was part of a natural cycle one would expect a natural reversal as well, but global change may not be reversible on the same time scales.

It is apparent that PDO cycles do exert a substantial influence on the

[101] S. Minobe, *Geophys. Res. Lett.*, 1999, **24**, 683–686.
[102] K. Odate, *Bull. Tohoku Nat. Fish. Res. Inst.*, 1994, **56**, 115–173.
[103] P. Muck, in D. Pauly, P. Muck, J. Mendo and I. Tsukayama (eds.), *The Peruvian Upwelling System: Dynamics and Interactions*, 1989, ICLARM Conference Proceedings, **18**, 386–403.

productivity of the Pacific Ocean, requiring adjustments in the food webs. It has been predicted that under current emission rates of greenhouse gases temperature increases would be large enough to shift the position of the thermal limits of sockeye salmon (*Oncorhynchus nerka*) by 2050.[104] Such a shift may exclude sockeye salmon from the entire Pacific Ocean, and severely restrict the overall area of the marine environment that would support growth. The impacts of such changes on the dynamics of the entire North Pacific Ocean and on the ecosystem's ability to adapt to natural PDO cycles remain to be seen.

3 Anthropogenic Effects

Climate/Fish Cycles and Industrial Fisheries

As we have noted through the case studies presented, basin-scale environmental conditions can vary substantially on decadal to multi-decadal timescales, regardless of observed long-term warming trends. At the global scale we also observed decadal cycles that may not be linked to global warming (Figure 2b). Global temperature oscillations with a period of 65–70 years, mainly in the Northern Hemisphere, have been observed.[105] Supported by results from 600-yr-long simulations of atmosphere–ocean general circulation models[106] it has been concluded that this oscillation arises from internal variability of the ocean–atmosphere system. Similar periodic oscillations have been observed in other atmospheric indices such as the length of day (reflecting the rate of spin of the Earth) and net air mass transport in the N/S and E/W directions (reflecting air pressure differences).[107]

Interestingly, multidecadal fluctuations in fish abundance have also been observed for several centuries, both during the pre-industrial[45,108] and industrial periods.[107,109,110] Figure 16 shows synchronized fluctuations in the catch trends of 10 of the most important commercial fish species in the 20th century.[107] The author concluded that most sardine (anchovy) species increase in phase with increases (decreases) in global temperature, suggesting that each species may favour particular climatic conditions. The relevance of these observations to global change research is that it suggests that fish fluctuations reflect changes in the biological productivity of the ocean, which may be atmospherically steered. If this is true then one should question the assumption of many global change models that the ocean is in a biological steady state.[111]

[104] D. W. Welch, B. R. Ward, B. D. Smith and J. P. Eveson, *Fish. Oceanogr.*, 2000, **9**, 17–32.

[105] M. E. Schlesinger and N. Ramankutty, *Nature*, 1994, **367**, 723–726.

[106] T. Delworth, S. Manabe and R. J. Stouffer, *J. Clim.*, 1993, **6**, 1993–2011.

[107] L. B. Klyashtorin, *Fish. Res.*, 1998, **37**, 115–125.

[108] T. R. Baumgartner, A. Soutar and V. Ferreira-Bartrina, *CALCOFI Rep.*, 1992, **33**, 24–40.

[109] D. H. Cushing, in B. J. Rothschild (ed.), *Toward a Theory on Biological Physical Interactions in the World Ocean*, Kluwer Academic Press, 1988, pp. 235–244.

[110] R. A. Schwartzlose, J. Alheit, A. Bakun, T. R. Baumgartner, R. Cloete, R. J. M. Crawford, W. J. Fletcher, Y. Green-Ruiz, E. Hagen, T. Kawasaki, D. Lluch-Belda, S. E. Lluch-Cota, A. D. MacCall, Y. Matsuura, M. O. Nevarez-Martinez, R. H. Parrish, C. Roy, R. Serra, K. V. Shust, M. N. Ward and J. Z. Zuzunaga, *S. Afr. J. Mar Sci.*, 1999, **21**, 289–347.

[111] B. J. Rothschild, in *Climate Change and Northern Fish Populations, Can. Sp. Publ. Fish. Aquat. Sci.*, 1995, **121**, 201–209.

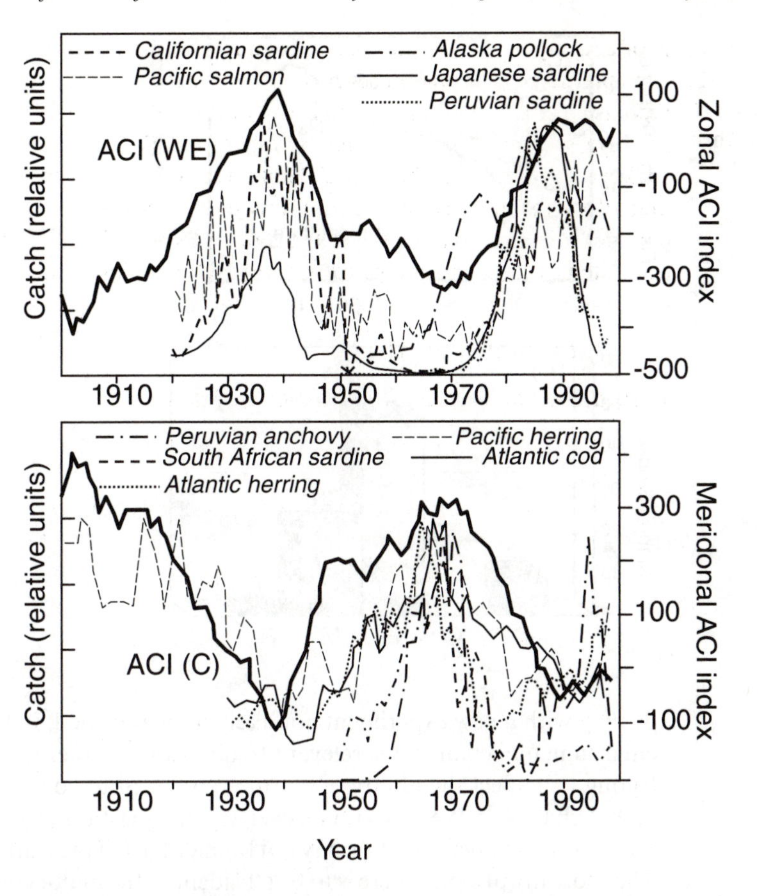

Figure 16 Catch trends in several major commercial species and dynamics of the zonal (WE) and meridional (WE) Atmospheric Circulation Index (ACI), a measure of hemispheric air mass transport. Meridional periods are related to global cooling, zonal periods are related to global warming[107]

The processes through which periodic variability of the ocean–atmosphere system affects biological production are largely unclear. One hypothesis, supported by concurrent declines in phytoplankton, zooplankton, herring and kittiwake stocks in the North Atlantic over a 30-year period,[73] is that marine ecological systems are bottom-up controlled. Wind intensity, through delays in the onset of thermal stratification, may be responsible for initiating this cascade. However, other hypotheses have been postulated,[112] each with distinctive responses to global change.

Ecosystem Effects of Fishing

Almost 100 million tonnes of fish are extracted from the sea every year (Figure 17a), providing 16% of the total animal protein for human consumption.[113] This figure has changed little in the last decade, and fluctuates slightly according to local phenomena such as El Niño events. The ecological consequences of extracting such a vast amount of protein from the sea cannot be dismissed. Because such exploitation is unparalleled in the history of the planet we are also

[112] J. G. Shepherd, J. G. Pope and R. D. Cousens, *Rapp. P.-v. Reun. Cons. Int. Explor. Mer*, 1984, **185**, 225–267.

[113] FAO, *State of the World Fisheries and Aquaculture*, 2000.

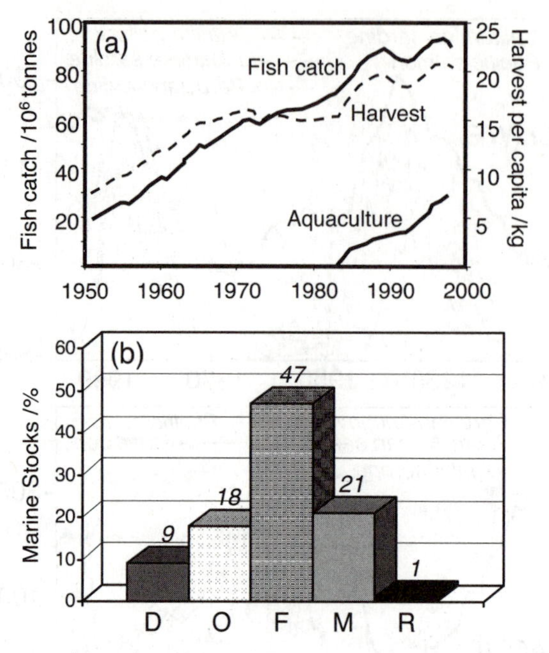

Figure 17 (a) World fish harvest 1950–1998 and (b) the state of marine fish stocks according to their exploitation levels:[113] D depleted, O over-exploited, F Fully exploited, M Moderately exploited, R Recovering

dealing with a new experiment not likely to be repeated. In the context of global environmental change it is relevant to question whether the ecosystem effects of fishing will affect the ability of marine ecosystem to cope with or/and adapt to global change.[82] We have no answer to this question beyond the limited logic that a system under pressure may not bounce back if pressure is further increased. Therefore my objective here is to briefly identify the major concerns regarding the consequences of fishing from an ecosystem perspective.

Commercial fishing tends to focus on a species or group of species (target species), but because fishing gears are not selective enough these are accompanied in the catch by a number of secondary species, or by-catch. Most by-catch is returned to the sea for economic, legal or personal reasons, largely in the form of dead fish. It has been estimated that global discards are in the order of 17.9 and 39.5 million tonnes per year,[115] a figure that varies heavily between fisheries (Table 4). Shrimp and prawn fisheries discard up to 89% of their total catch while pelagic fisheries like mackerel and sardine do not discard more than 10% of the total catch, although this still equates to over a million tonnes of fish per year. A substantial proportion of these discards, perhaps over 75%, is not lost, being consumed by higher predators like seabirds and mammals.[116] By-catch is arguably the biggest problem faced by the world's fisheries, both from a management point of view and from an ecological point of view.

Like excessive by-catch, intensive fishing also changes the structure of communities if the frequency of extraction is shorter than the generation time of

[114] D. Pauly and V. Christensen, *Nature*, 1995, **374**, 255–257.

[115] D. L. Alverson, in *Global Trends: Fisheries Management*, eds. E. K. Pikitch, D. D. Huppert and M. P. Sissenwine, 1997, Amer. Fisheries Society Symposium, **20**, MD, USA.

[116] A. V. Hudson and R. W. Furness, *J. Zool.*, 1988, **215**, 151–166.

Table 4 Global estimates of primary production and proportion of primary production required to sustain global fish catches[114]

Ecosystem	*Area/* $10^6\,km^2$	*Primary production/* $g\,C\,m^{-2}\,yr^{-1}$	*Catch/* $g\,m^{-2}\,yr^{-1}$	*Discards/* $g\,m^{-2}\,yr^{-1}$	*Mean % of primary production*
Open ocean	332.0	103	0.01	0.002	1.8
Upwellings	0.8	973	22.2	3.36	25.1
Tropical shelves	8.6	310	2.2	0.671	24.2
Non-tropical shelves	18.4	310	1.6	0.706	35.3
Coastal reefs	2.0	890	8.0	2.51	8.3

some of the species involved. In one of the most illustrative examples fishing effort, gear efficiency and the initial composition of a benthic community in the Southeastern North Sea were used to estimate the population trajectories of the main species from 1945 to 1983.[117] In less than 40 years the area changed from being dominated by fish like Greater Weevers and Rokers to invertebrates like swimming crabs, shells and anemones. The significance of these changes in the context of global change is that such dramatic turnabouts would have important consequences on the expected ecosystem responses to physical forcing. For example, predators often regulate the transfer of energy between trophic levels and thus control the food webs of some ecosystems (*e.g.* Baltic Sea[118]). The disappearance of these top predators is likely to influence the responses of the whole ecosystem to, for example, increased water temperatures, as the controllers would change as well. In order to understand this better we should look at measures of community structure and how these measures have changed over the period of fisheries industrialization.

Species diversity is a measure of community structure that has been used to indicate ecosystem resilience. Studying species diversity trends during the development and establishment of fisheries in the Georges Bank and in the Gulf of Thailand it was observed that diversity increased during early stages of the fishery, and then declined[119] (Figure 18). As global change includes short-term anthropogenic pressures like overfishing and changing fishing practices, we can say that global change has radically affected the structure of these ecosystems by exerting a top-down control over them.

However, diversity does not always tell the whole story, as some heavily exploited ecosystems do not show clear trends following industrialization.[122] Some studies have shown how diversity may not be reduced as a result of heavy fishing pressures, but fish size has.[123] This was reflected in a recent study of the mean trophic level of the catches in the Northwest Atlantic, 1950–1994.[124] Trophic level reflects not only the mean size of the animals, but also quantifies the

[117] J.C.M. Philippart, *ICES J. Mar. Sci.*, 1997, **55**, 342–352.
[118] I. Rudstam, G. Aneer and M. Hilden, *Dana*, 1994, **10**, 105–129.
[119] S.J. Hall, *Fish Biol. Aquat. Resour. Ser.*, 1999, **1**, 274 pp.
[120] D. Pauly, *RAPA Report*, 1987, **1987/10**.
[121] A.R. Solow, *Biometrics*, 1994, **50**, 556–565.
[122] S.P.R. Greenstreet and S.J. Hall, *J. Anim. Ecol.*, 1996, **65**, 577–598.
[123] J.G. Pope and B.J. Knights, in M.C. Mercer (ed.), *Ecodynamics: Contributions to Theoretical Ecology, Can. Sp. Publ. Fish. Aquat. Sci.*, 1982, pp. 116–118.
[124] D. Pauly, V. Christensen, J. Dalsgaard, R. Forese and F. Torres, *Science*, 1998, **279**, 860–863.

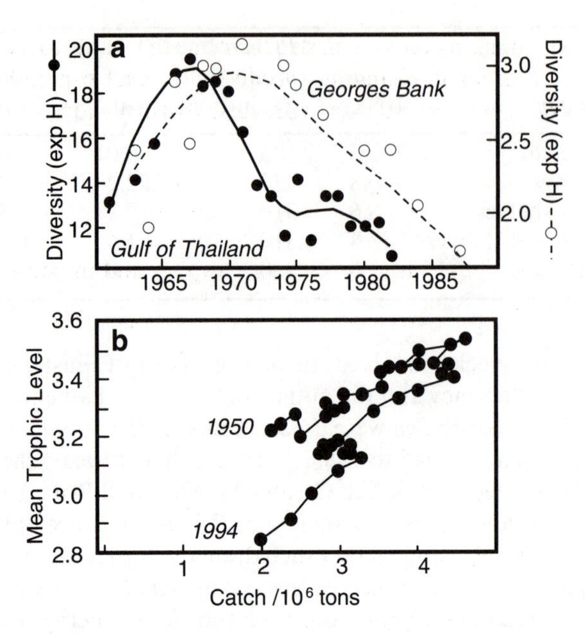

Figure 18 Anthropogenic changes in marine ecosystem structure. (a) Trends in the Shannon–Wiener diversity index (H) in the Gulf of Thailand[120] and Georges Bank[121] and (b) mean trophic level *versus* catch in the Northwest Atlantic, 1950–1994[124]

replacement of top predators in favour of lower predators.[117] In the Northwest Atlantic the mean trophic level of the catches increased in the early stages of the fishery (Figure 18b), and then declined. The authors also argue that the results negate the hypothesis that fishing at lower trophic levels will give greater yields by reducing the amount of energy losses. For our purposes it is enough to state that heavy fishing is severely affecting the structure of many marine ecosystems, generally the most productive ones. This structure has been reached over long periods of time based on complex multispecific interactions, including trophic links, density-dependent responses and competition for space and food. Although the matter is largely speculative, it seems intuitively correct to expect that the responses of these ecosystems to physical forcing and global change will be dramatically affected by these structural changes.

4 Conclusions and Way Forward

In this review I have tried to show how concerns about the impacts of climate variability and anthropogenic forcing on marine ecosystems are global. Understanding these impacts requires insight into a wide range of fluctuations and oscillations that range from seasonal to multi-decadal. In the past decade the dominant focus of concern in global change research has been on changes from a single climatic parameter (*e.g.* atmospheric temperature) on single processes, as if these could be isolated. In the future we will have to focus on the interactions and feedbacks among agents and impacts of change at the species, ecosystem, basin and global scales. Coral reefs may be a good example of the nature of multiple and interactive stresses. Natural disturbances, like hurricanes, natural bleaching and diseases are all part of coral reef dynamics. Global change has added a number of disturbances over recent years, from increasing nutrient loadings from on-shore activities, changing ecosystem structure due to fishing, to increases in atmospheric CO_2. The latter is changing the ability of reef organisms to create calcium carbonate shells. Global warming is also causing widespread bleaching[71,125] but the fate of corals will not be determined by a single cause–effect relationship, but rather by the interactive relationship between a number of human-driven stresses. Similar analogies could be drawn throughout the world's oceans.

Feedback processes also need to be considered. For example, during El Niño events coastal upwelling is severely reduced, and in consequence the release of CO_2 from the equatorial Pacific to the atmosphere is constrained. If the frequency of El Niño events increases as a result of global warming this mechanism may provide a negative feedback loop, reducing the release of greenhouse gases to the atmosphere. However, a warmer ocean would have less capacity to dissolve atmospheric CO_2, providing a positive feedback loop. In order to evaluate the magnitude of these feedback mechanisms we must identify the principal pathways of energy and materials in key ecosystems, and evaluate how these fluctuate as a result of global change.

Humans will be the ultimate receivers of changes in marine ecosystem structure. An obvious direct concern would be the need to re-evaluate the use of our marine resources should climate change and our own activities threaten the supply of food. FAO projections for 2100 suggest that the amount of fish protein per capita would at best be similar to 1998 levels, and at worse about a third lower. Global change can reduce this even further. Improving our use of the 25% of the total fish catch that is discarded annually as by-catch is a priority, as may be the use of the landed catch. In this regard about 25% of the total fish landed is used for animal feeds, often inefficiently. For example the proportion of fish meal supplies used for farming fish has risen from 10% in 1988 to 33% in 1997, both reflecting a trend towards farming carnivorous fish as well as the desire to increase the growth of non-carnivorous species. Such trends bring into question whether in the future fish farming will actually add to world fish supplies, the very reason why aquaculture has been encouraged to develop over the last decades.[126] Fishing down pelagic resources to satisfy the expensive demands for top foods of the developed world may be a faster way of generating further changes in marine ecosystem structure than waiting for climate change to act.

The challenge of understanding a changing Earth demands the development of a substantive science of integration. This science should be built on complex systems analysis that addresses the synergies, interactions and non-linearities that defy traditional cause–effect relationships. It must transcend disciplinary boundaries across natural and possibly social science. The tools of the game are available: paleo-environmental research, to understand processes that operate on long time scales; Monitoring and observation systems, to make the most of the expanding array of sophisticated remote sensors; Process studies, aimed at resolving hot spots and bottom-up processes; and simulations, to describe and

[125] P. W. Glynn, *Tree*, 1991, **6**, 175–179.

[126] R. L. Naylor, R. J. Goldburg, J. H. Primavera, N. Kautsky, M. C. M. Beveridge, J. Clay, C. Folke, J. Lubchenco, H. Mooney and M. Tipell, *Nature*, 2000, **405**, 1017–1024.

understand the Earth's switch and choke points. Through this science we can face the challenge of ensuring sustainable life support systems and developing an integrated management of the Earth's system.

5 Acknowledgement

I wish to thank Dr Roger Harris for his comments on an early version of this article.

Rising Sea Levels: Potential Impacts and Responses

ROBERT J. NICHOLLS

1. Introduction

Global-mean sea levels are expected to rise through the 21st century and beyond due to human-induced global warming. Given the large population in the coastal zone, these changes have important long-term implications for societal use of the coastal zone. Figure 1 suggests that about 20 and 37% of global population live within 30 and 100 km from the coast, respectively (see caveats of Small *et al.*[1]). A considerable portion of global GDP is also produced in coastal zones.[2,3] In addition, it is widely reported that many coastal locations are experiencing population and GDP growth that is higher than their national averages,[4,5] and there is a significant urbanizing trend.[6] This suggests a net coastward migration of people and economic activities. Thus, human exposure to the effects of sea-level rise and climate change is growing rapidly.

Coastal areas are already threatened by a range of coastal hazards such as erosion, saline intrusion, flooding and tsunamis: all will be exacerbated by sea-level rise. Additionally, human activity is producing profound changes to the world's coastal zone such as reducing freshwater inputs, modifying sediment budgets, armouring the coast and direct and indirect ecosystem destruction.[7] Therefore, climate change and sea-level rise are an *additional* pressure on

[1] C. Small, V. Gornitz and J. Cohen, Coastal hazards and the global distribution of human population, *Environ. Geosci.*, 2000, **7**, 3–12.

[2] R.K. Turner, S. Subak and W.N. Adger, Pressures, trends, and impacts in coastal zones: interactions between socioeconomic and natural systems. *Environ. Manage.*, 1996, **20**(2), 159–173.

[3] J.D. Sachs, A.D. Mellinger and J.L. Gallup, The geography of poverty and wealth, *Sci. Am.*, 2001, **284**(3), 70–75.

[4] WCC'93, *Preparing to Meet the Coastal Challenges of the 21st Century*. Report of the World Coast Conference, Noordwijk, The Netherlands, 1–5 November 1993, Ministry of Transport, Public Works and Water Management, The Hague, The Netherlands, 1994, 49s pp. + apps.

[5] R. Hausmann, 2001, Prisoners of geography, *Foreign Policy*, 2001, No 122 (January/February), 45–53.

[6] R.J. Nicholls, Coastal megacities and climate change, *Geojournal*, 1995, **37**, 369–379.

[7] P. Holligan and H. deBoois (eds.), *Land-Ocean Interactions in the Coastal Zone (LOICZ). Science Plan.* International Geosphere Biosphere Programme, International Council of Scientific Unions, Stockholm, 1993, 50 pp.

Issues in Environmental Science and Technology, No. 17
Global Environmental Change

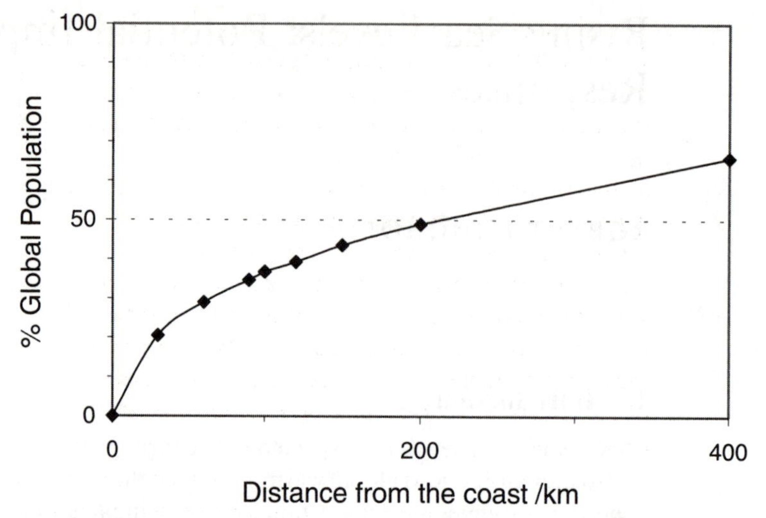

Figure 1 Population *versus* distance from the coast estimated for 1994[8,9] based on analysis of the Gridded Population of the World Dataset, Version 1[10]

coastal areas which will interact adversely with these other trends.[11,12]

This article examines the likely implications of climate change for coastal areas in the 21st century, with a main emphasis on sea-level rise, and how we can systematically analyse the potential impacts and response options. First, sea-level rise and climate change in coastal areas are considered to indicate the likely magnitude of change. Second, impact and vulnerability analyses are considered, including some examples of potential impacts. Third, possible responses are considered, with an emphasis on adaptation to sea-level rise and climate change. Last, the future needs to improve impact and vulnerability assessment are considered.

2 Sea-level Rise and Climate Change in Coastal Areas

Sea-level Rise

Sea levels have fluctuated throughout geological time, including a major global

[8] J. E. Cohen, C. Small, A. Mellinger, J. Gallup and J. Sachs, Estimates of coastal populations, *Science*, 1997, **278**, 1211–1212.

[9] R. Gommes, J. du Guerny, F. Nachtergaele and R. Brinkman, *Potential Impacts of Sea-Level Rise on Populations and Agriculture.* SD-Dimensions Special, FAO, Rome, 1998, http://www.fao.org/sd/eidirect/EIre0045.htm.

[10] W. Tobler, U. Deichmann, J. Gottsegen and K. Malloy, K., World population in a grid of spherical quadrilaterals. *Int. J. Population Geogr.*, 1997, **3**, 203–225 (see also http://sedac.ciesin.org/plue/gpw).

[11] R. McLean, A. Tsyban, V. Burkett, J. O. Codignotto, D. L. Forbes, N. Mimura, R. J. Beamish and V. Ittekkot, Coastal zone and marine ecosystems, in J. J. McCarthy, O. F. Canziani, N. A. Leary, D. J. Dokken and K. S. White (eds.), *Climate Change 2001: Impacts, Adaptation and Vulnerability*, Cambridge University Press, Cambridge, 2001, pp. 343–380.

[12] L. Bijlsma, C. N. Ehler, R. J. T. Klein, S. M. Kulshrestha, R. F. McLean, N. Mimura, R. J. Nicholls, L. A. Nurse, H. Pérez Nieto, E. Z. Stakhiv, R. K. Turner and R. A. Warrick, Coastal zones and small islands, in R. T. Watson, M. C. Zinyowera and R. H. Moss (eds.), *Climate Change 1995 – Impacts, Adaptations and Mitigation of Climate Change: Scientific-Technical Analyses*, Cambridge University Press, Cambridge, 1996, pp. 289–324.

rise in sea level of over 120 m that occurred from 15 000 to 6000 years ago.[13,14] This was related to significant global warming and melting of large ice caps situated in the northern hemisphere. Now, human-induced climate change threatens a renewed global rise in sea level.[14]

Sea-level components. The local change in sea level at any coastal location depends on the sum of global, regional and local factors and is termed relative sea-level change.[15] Therefore, the global-mean sea-level rise does not translate into a uniform rise in sea level around the world. The relative (or local) level of the sea to the land can change for a number of reasons and over a range of timescales from seconds to millions of years.[16,17] Here we are interested in mean sea level, so the time period of wave and tides can be disregarded, while the upper time bound is 10^2–10^3 years. Over this period, relative sea level is the sum of several components:[14]

- *Global-mean sea-level rise*, which is due to an increase in the global volume of the ocean. In the 20th/21st century, this is primarily due to thermal expansion of upper ocean as it warms and the melting of small ice caps due to human-induced global warming.[14] The contribution of Greenland is less certain, while the Antarctica ice sheet is expected to grow in size due to increased snowfall. This will produce a relative sea-level *fall*, offsetting any positive contribution from Greenland.[14] Beyond 2100, Greenland and Antarctica could become significant contributors to sea-level rise. Direct human influence is also possible due to modifications to the hydrological cycle [*e.g.* increased terrestrial storage of water (causing sea-level fall), *versus* increased groundwater mining (causing sea-level rise)].
- *Regional meteo-oceanographic factors* such as spatial variation in thermal expansion effects, changes to long-term wind fields and atmospheric pressure, and changes in ocean circulation such as the Gulf Stream.[18] These effects could be significant, with regional effects equal to the magnitude of the global-mean thermal expansion term. A possible example of this type of effect is an apparent fall of the Mediterranean since 1960.[19] This observation might be linked to decreased freshwater inputs to the Mediterranean and

[13] M. S. Kearney, Late holocene sea level variations, in B. C. Douglas, M. S. Kearney and S. P. Leatherman (eds.), *Sea-level Rise: History and Consequences*, Academic Press, London, 2000, pp. 13–36.

[14] J. A. Church, J. M. Gregory, P. Huybrechts, M. Kuhn, K. Lambeck, M. T. Nhuan, D. Qin and P. L. Woodworth, Changes in sea level, in D. J. Houghton, Y. Ding, D. J. Griggs, M. Noguer, P. J. van der Linden and D. Xiaosu (eds.), *Climate Change 2001. The Scientific Basis*, Cambridge University Press, Cambridge, 2001, pp. 639–693.

[15] R. J. Nicholls and S. P. Leatherman, Adapting to sea-level rise: relative sea level trends to 2100 for the USA, *Coastal Manage.*, 1996, **24**(4), 301–324.

[16] R. J. N. Devoy, *Sea Surface Studies: A Global View*, Croom Helm, London, 1987, 650 pp.

[17] D. Smith, S. B. Raper, S. Zerbini and A. Sanchez-Arcilla, (eds.), *Sea Level Change and Coastal Processes. Implications for Europe*, EUR 19337, European Commission, Brussels, 2000, 247 pp.

[18] J. M. Gregory, Sea-level changes under increasing atmospheric CO_2 in a transient coupled ocean-atmosphere GCM experiment, *J. Clim.*, 1993, **6**, 2247–2262.

[19] M. N. Tsimplis and T. F. Baker, Sea level drop in the Mediterranean Sea: an indicator of deep water salinity and temperature changes?, *Geophys. Res. Lett.*, 2000, **27**, 1731–1734.

hence has a direct human influence. Models of these effects under global warming show little agreement,[20] so detailed scenarios are not yet possible and this component has been ignored in impact assessments to date.

- *Vertical land movement* (subsidence/uplift) due to various geological processes such as tectonics, neotectonics, glacial-isostatic adjustment (GIA), and consolidation.[21] While the earth may appear stable, vertical land movement is almost universal, although the rate of change varies significantly. During the 20th century, land movements due to GIA appear to have been comparable to global-mean sea-level rise.[22] (GIA is still occurring due to the unloading of melting ice sheets from 18 000 to 6000 years ago). In addition to natural changes, groundwater withdrawal and improved drainage has enhanced subsidence (and peat destruction by oxidation and erosion) in many coastal lowlands, producing several metres subsidence in susceptible areas over the 20th century, including some major coastal cities.[6]

Historic sea-level trends. Global sea levels are estimated to have risen 10–20 cm during the 20th century with no evidence of acceleration.[14] Figure 2 shows selected relative sea-level records from Europe for this period, measured with tide gauges. They show variable rising trends of $\leqslant 1.6\,\mathrm{mm\,yr^{-1}}$, except Helsinki where relative sea level is falling at $2.5\,\mathrm{mm\,yr^{-1}}$ due to GIA-induced uplift, which more than compensates for the global-mean rise. Aberdeen is also experiencing uplift, while elsewhere relative sea-level rise is within the range of the global-mean rise. Interestingly, Mitrovica *et al.*[23] have suggested that sea-level rise in much of Europe is being suppressed by GIA-induced uplift in response to *contemporary* melting of Greenland. While this remains a preliminary result which requires further investigation, it illustrates the large number of factors that need to be considered when evaluating observations and future relative sea levels.

While there is no evidence of an acceleration in sea-level rise during the 20th century, the limited sea-level measurements from the 18th and 19th century do show an acceleration,[14,24] and geological data provide further support for this conclusion.[25,26] The important implication is that coastal environments around the world are already experiencing rates of sea-level rise which exceed those over

[20] J. M. Gregory, J. A. Church, K. W. Dixon, G. M. Flato, D. R. Jackett, J. A. Lowe, J. M. Oberhuber, S. P. O'Farrell and R. J. Stouffer, Comparison of results from several AOGCMs on global and regional sea level change 1900–2100, *Climate. Dynamics*, in press.

[21] K. O. Emery and D. G. Aubrey, *Sea Levels, Land Levels and Tide Gauges*, Springer Verlag, New York, 1991.

[22] B. C. Douglas, Sea-level Change in the Era of the Recording Tide Gauge, in B. C. Douglas, M. S. Kearney and S. P. Leatherman (eds.), *Sea-level Rise: History and Consequences.* Academic Press, London, 2000, pp. 37–64.

[23] J. X. Mitrovica, M. E. Tamisiea, J. L. Davis, and G. A. Milne, Recent mass balance of polar ice sheets inferred from patterns of global sea-level change, *Nature*, 2001, **409**, 1026–1029.

[24] P. L. Woodworth, High waters at Liverpool since 1768: the UK's longest sea-level record, *Geophys. Res. Lett.*, 1999, **26**, 1589–1592.

[25] I. Shennan and P. L. Woodworth, A comparison of late holocene and twentieth-century sea-level trends from the UK and North Sea region, *Geophys. J. Int.*, 1992, **109**, 96–105.

[26] A. Long, Late holocene sea-level change, *Prog. Phys. Geogr.*, 2000, **24**, 415–423.

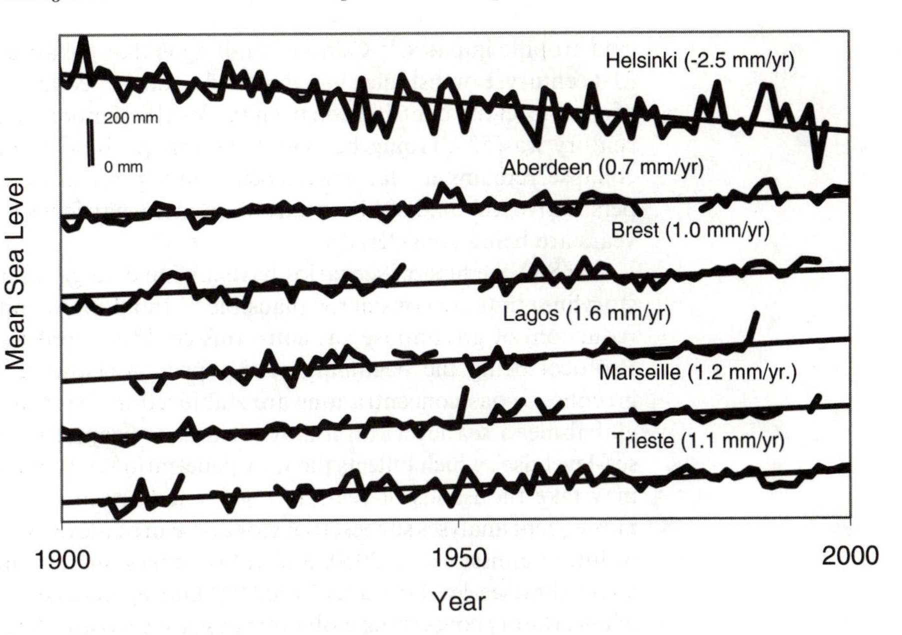

Figure 2 Selected European mean sea-level records for the 20th century, including the linear trend: Trieste, Italy; Marseille, France; Lagos, Portugal; Brest, France; Aberdeen, UK; and Helsinki, Finland. The offsets between records are for display purposes. Data from the Permanent Service for Mean Sea Level (PSMSL; see web site http://www.pol.ac.uk/psmsl/)

the last several thousand years, while many coasts that were experiencing stable or falling sea levels are now experiencing sea-level rise.[27]

Future sea-level scenarios. Global-mean sea-level rise is expected to accelerate during the 21st century due to human-induced global warming. Taking the new emission scenarios from the Special Report on Emission Scenarios (SRES),[28] the Intergovernmental Panel on Climate Change (IPCC) estimate that the global rise from 1990 to 2100 will be between 9 and 88 cm, with a mid estimate of 48 cm.[14] This is slightly lower than the previous IPCC assessment,[29] but the large range of uncertainty for global-mean rise remains.

These scenarios do not include the possible collapse of the West Antarctic Ice Sheet (WAIS), which could abruptly raise global sea levels by up to 6 m,[30] causing

[27] I. Shennan, K. Lambeck, B. Horton, J. Innes, J. Lloyd, J. McArthur and M. Rutherford, Holocene isostacy and relative sea-level changes on the east coast of England, in I. Shennan, and J. Andrews, (eds.), *Holocene Land-Ocean Interaction and Environmental Change around the North Sea*, Geological Society, London, Special Publications 166, 1999, pp. 275–298.

[28] N. Nakicenovic, J. Alcamo, G. Davis, B. de Vries, J. Fenhann, S. Gaffin, K. Gregory, A. Grubler, T. Y. Jung, T. Kram, E. L. La Rovere, L. Michaelis, S. Mori, T. Morita, W. Pepper, H. Pitcher, L. Price, K. Riahi, A. Roehrl, H.-H. Rogner, A. Sankovski, M. Schlesinger, P. Shukla, S. Smith, R. Swart, S. van Rooijen, N. Victor and Z. Dadi, *Special Report on Emissions Scenarios: A Special Report of Working Group III of the Intergovernmental Panel on Climate Change*, Cambridge University Press, Cambridge, 2000, 599 pp.

[29] R. A. Warrick, C. Le Provost, M. F. Meier, J. Oerlemans and P. L. Woodworth, Changes in sea level, in J. T. Houghton, L. G. Meira Filho, B. A. Callander, N. Harris, A. Kattenberg and K. Maskell (eds.), *Climate Change 1995 – The Science of Climate Change*, Cambridge University Press, Cambridge, 1996, pp. 359–405.

[30] J. H. Mercer, West Antartic ice sheet and CO_2 greenhouse effect: a threat of disaster. *Nature*, 1978, **271**, 321–325.

catastrophic impacts.[31] Current thinking is that collapse is unlikely during the 21st century. For instance, Vaughan and Sponge[32] concluded that the probability of a sea-level rise contribution from the WAIS of more than 0.5 m during the 21st century was 5%. Going beyond 2100, this probability increases. Thus, WAIS collapse remains a plausible, albeit unlikely scenario, and from an impacts perspective it should not be totally discounted, particularly if changes over 100+ years are being considered.

The SRES emission scenarios have a wide divergence, reflecting the different storylines that are considered plausible.[28] In addition, mitigation (or deliberate reduction) of greenhouse gas emissions could be implemented with the Kyoto Protocol being the beginning of this process. However, even if atmospheric greenhouse-gas concentrations are stabilized in the next few decades, a rise in global-mean sea level still follows. This has been termed 'the commitment to sea-level rise', which reflects the slow penetration of heat into the deeper ocean: it may take thousands of years to reach equilibrium with the new conditions.[33] Subsequent analyses suggest that global-mean sea-level rise is almost independent of future emissions to 2050, and future emissions become most important in controlling sea-level rise after 2100.[14,34] During the 21st century, the main source of uncertainty concerning global-mean sea-level rise is the climate response to the emissions, particularly the climate sensitivity. Therefore, some global-mean sea-level rise appears inevitable during the 21st century and beyond, even given substantial mitigation of climate change. (Note the important point that mitigation also makes abrupt climate change events such as the WAIS collapse even more unlikely. This benefit of mitigation is almost always ignored.)

An example of relative sea-level observations and a high, mid and low scenario based on the scenarios of Church *et al.* are shown for New York City in Figure 3. The rise during the 20th century (30 cm/century) is 10–20 cm/century larger than the global-mean trend,[14] reflecting the fact that New York is slowly subsiding due to GIA.[35] The relative mid scenario shows a 2-fold acceleration relative to the 20th century. However, the range is more than a 3-fold acceleration to a slight deceleration. Note that the scenarios assume no meteo-oceanographic effects at New York, but they do take account of the uncertainty in subsidence. All these uncertainties need to be considered when assessing responses to sea-level rise (see later).

[31] S. H. Schneider and R. S. Chen, Carbon dioxide flooding: physical factors and climatic impact, *Annu. Rev. Energy*, 1980, **5**, 107–140.

[32] Referenced in J. A. Church, J. M. Gregory, P. Huybrechts, M. Kuhn, K. Lambeck, M. T. Nhuan, D. Qin and P. L. Woodworth, Changes in sea level, in J. T. Houghton, Y. Ding, D. J. Griggs, M. Noguer, P. J. van der Linden and D. Xiaosu (eds.), *Climate Change 2001. The Scientific Basis*, Cambridge University Press, Cambridge, 2001, pp. 639–693.

[33] T. M. L Wigley and S. C. B. Raper, Future changes in global mean temperature and sea level, in R. A. Warrick, E. M. Barrow and T. M. L. Wigley (eds.), *Climate and Sea Level Change: Observation, Projections and Implications*, Cambridge University Press, Cambridge, 1993, pp. 111–133.

[34] T. M. L. Wigley, *The Science of Climate Change: Global and U.S. Perspectives*, Report for the Pew Center on Global Climate Change, Arlington, VA, 1999, 51 pp. (http://www.pewclimate.org/projects/env—science.cfm)

[35] B. C. Douglas, Global sea-level rise. *J. Geophys. Res.*, 1991, **96** (C4), 6981–6992.

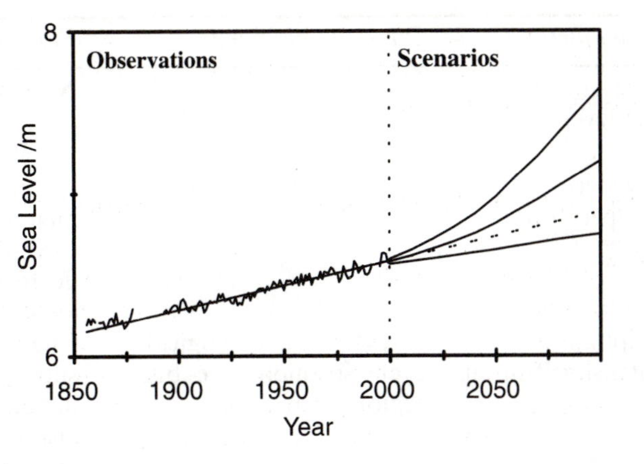

Figure 3 Sea level at New York City from 1850 to 2100. Relative sea-level rise observations combined with relative sea-level rise scenarios derived from Church *et al.*[14] The straight dashed line shows the observed trend during the 20th century

Other Climate Change

Other aspects of coastal climate are likely to change under global warming with many adverse and some beneficial effects (Table 1). The influence of climate change on the long-term tracks, intensity and frequency of coastal storms is of widespread concern due to the high impact potential.[14,36,37] This factor receives considerable attention and erroneous comments about the certainty of an increase in storminess under global warming are also common. Given the high interannual and interdecadal variability of storm occurrence[38,39] it may be difficult to discern any long-term changes from natural variability. In this regard, new measurement techniques may be helpful.[40]

Following the uncertainties about other climate change factors, the main focus of most assessments has been the impacts and responses to sea-level rise. This article follows a similar approach.

3 Impact and Vulnerability Assessment of Sea-level Rise

Impact assessment provides important insights into a coastal system's vulnerability to sea-level rise and climate. In general terms, *vulnerability* is 'the degree to which a system is susceptible to injury, damage, or harm' from whatever cause, and is a product of a system's *sensitivity* and *adaptive capacity* to the stress.[41] Vulnerability

[36] A. B. Pittock and R. A. Flather, Severe tropical storms and storm surges, in R. A. Warrick, E. M. Barrow and T. M. L. Wigley (eds.), *Climate Change and Sea Level Change: Observations, Projections and Implications*, Cambridge University Press, Cambridge, 1993, pp. 393–394.

[37] R. A. Warrick, K. L. McInnes, A. B. Pittock and P. S. Kench, Climate change, severe storms and sea level: implications for the coast, in D. J. Parker (ed.), *Floods*, Volume 2, Routledge, London, 2000, pp. 130–147.

[38] WASA Group, Changing waves and storms in the Northeast Atlantic, *Bull. Am. Meteorol. Soc.*, **79**, 1998, 741–760.

[39] K. Zhang, B. C. Douglas and S. P. Leatherman, Twentieth-century storm activity along the US east coast, *J. Clim.*, 2000, **13**, 1748–1761.

[40] I. Grevemeyer, R. Herber and H.-H. Essen, Microseismological evidence for a changing wave climate in the northeast Atlantic Ocean, *Nature*, 2000, **408**, 349–352.

Table 1 Some climate change and related factors relevant to coasts and their biogeophysical effects

Climate factor	*Direction of change*	*Biogeophysical effects*
Global-mean sea level	+ve	Numerous (see Table 2)
Sea water temperature	+ve	Increased coral bleaching; migration of coastal species towards higher latitudes; decreased incidence of sea ice at higher latitudes
Precipitation intensity/Run-off	Intensified hydrological cycle, so often +ve, but regional variation	Changed fluvial sediment supply; changed flood risk in coastal lowlands; but also consider catchment management
Wave climate	Poorly known, but significant temporal and spatial variability expected	Changed patterns of erosion and accretion; changed storm impacts
Storm track, frequency and intensity	Poorly known, but significant temporal and spatial variability expected	Changed occurrence of storm flooding and storm damage
Atmospheric CO_2	+ve	Increased productivity in coastal ecosystems; decreased $CaCO_3$ saturation impacts on coral reefs

of coastal zones to climate change and sea-level rise has been defined as 'the degree of incapability to cope with the consequences' of these stresses.[42] Thus, vulnerability assessment includes the assessment of both anticipated impacts and available adaptation options.

Our understanding of the impacts of sea-level rise is often conditioned by the biogeophysical (or natural) system response to rapid sea-level rise during the early Holocene. However, the coastal system can now be characterized as an evolving, coupled natural-human system.[43] Therefore, the present and future consequences of sea-level rise need to be analysed and interpreted with these changed conditions in mind. As discussed below, sea-level rise has a variety of

[41] B. Smit, O. Pilifosova, I. Burtin, B. Challenger, S. Huq, R. J. T. Klein and G. Yohe, Adaptation to climate change in the context of sustainable development and equity, in J. J. McCarthy, O. F. Canziani, N. A. Leary, D. J. Dokken and K. S. White (eds.), *Climate Change 2001: Impacts, Adaptation and Vulnerability*, Cambridge University Press, Cambridge, 2001, pp. 877–912.

[42] IPCC CZMS, A common methodology for assessing vulnerability to sea-level rise – second revision, in *Global Climate Change and the Rising Challenge of the Sea*, Report of the Coastal Zone Management Subgroup, Response Strategies Working Group of the Intergovernmental Panel on Climate Change, Ministry of Transport, Public Works and Water Management, The Hague, The Netherlands, 1992, Appendix C, 27 pp.

[43] R. J. T. Klein and R. J. Nicholls, Assessment of coastal vulnerability to climate change, *Ambio*, 1999, **28** (2), 182–187.

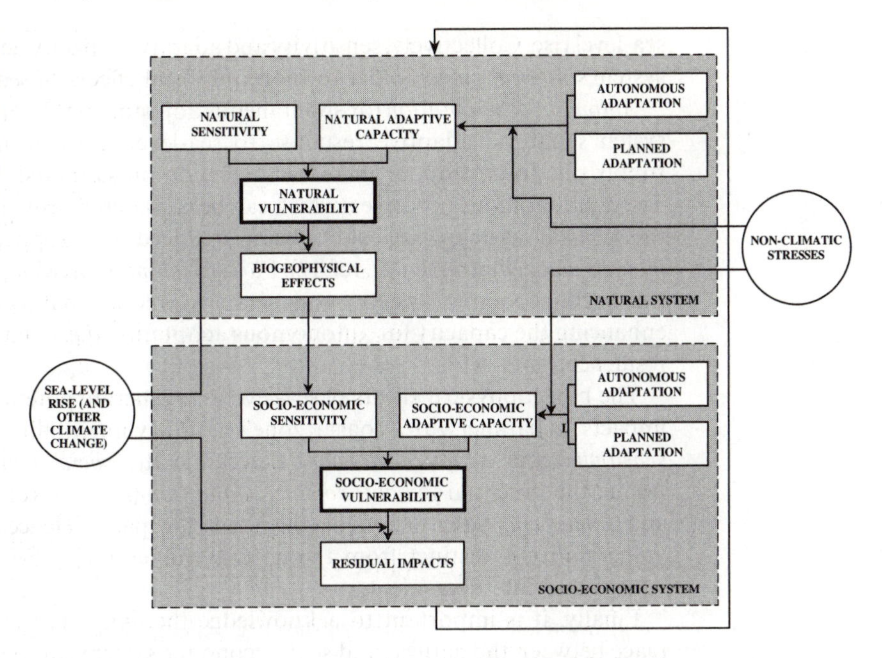

Figure 4 A conceptual framework for coastal impact and vulnerability assessment of sea-level rise (adapted from Klein and Nicholls[43,44] to reflect terminology in the IPCC Third Assessment Report[41]

biogeophysical effects and socio-economic impacts. For any single effect or impact, there is also a variety of methods and approaches to analysis, leading to a wide range of approaches which could be applied to assessing the overall impacts of sea-level rise. Last, the interactions between the different effects and impacts need to be considered, including the scope for adaptation. Therefore, it is critical to have:

- a conceptual framework within which to ask basic questions about impacts and vulnerability;
- a consistent vulnerability assessment methodology to address these questions; and
- appropriate analytical tools to implement the methodology.

Thus, the conceptual framework is overarching and key to the analysis: an example is presented in Figure 4. The terminology has been modified slightly from the original to reflect the terms used by Smit *et al.*,[41] but the underlying meanings remain the same.

Figure 4 distinguishes the natural system and socio-economic system which coexist in the coastal zone. Analysis of socio-economic impacts and vulnerability requires prior understanding of the natural system response to sea-level rise. Hence, analysis of coastal vulnerability starts with some notion of the *natural(-system) sensitivity* to the biogeophysical effects of sea-level rise, and the *natural(-system) adaptive capacity* to cope with these effects. Sensitivity simply reflects the natural system's potential to be affected by sea-level rise (*e.g.* a subsiding sedimentary delta is more sensitive than an emerging rocky coast), while adaptive capacity describes the natural system's stability in the face of sea-level rise. Collectively, sensitivity and adaptive capacity determine the coastal system's *natural vulnerability* to biogeophysical effects of sea-level rise.

Autonomous adaptation (or spontaneous adjustments[45]) represents the coastal system's natural adaptive response to sea-level rise (*e.g.* increased sediment supply due to erosion, or increased accretion on saltmarshes and mangroves). These autonomous processes (which can be considered resilient characteristics of the coastal system) are often being reduced or stopped by *non-climatic (human-induced) stresses.* (*cf.* ref. 12). *Planned adaptation* (which must emerge from the socio-economic system) can serve to reduce natural vulnerability by enhancing the capacity for autonomous adaptation (*i.e.* enhance natural-system resilience).

The biogeophysical effects of sea-level rise generate potential socio-economic impacts. In parallel with a coastal zone's natural vulnerability, the *socio-economic sensitivity* and *adaptive capacity* determine the *socio-economic vulnerability*. Again, the potential for *autonomous adaptation* (*e.g.* market price adjustments) and *planned adaptation* determines this adaptive capacity. Hence, the socio-economic vulnerability is distinct from the natural vulnerability, even though they are related and interdependent.

Finally, it is important to acknowledge the dynamic interaction that takes place between the natural and socio-economic systems in the coastal zone. This includes the natural system impacts on the socio-economic system and planned adaptation by the socio-economic system influencing the natural system. Hence, the natural and socio-economic systems interact in a complex manner which needs to be taken into account in any analysis. The biogeophysical effects and socio-economic impacts of sea-level rise are now considered within the framework outlined in Figure 4.

Biogeophysical Effects of Sea-Level Rise

Climate change and sea-level rise have a wide range of effects on coastal processes.[46] In addition to raising ocean level, rising sea level also raises all the coastal processes that operate around sea level. Therefore, the immediate effect of a rise in sea level concerns submergence and increased flooding of coastal land, and saltwater intrusion of surface waters. Longer-term effects, including morphological change and saltwater intrusion into groundwater as the coast adjusts to the new environmental conditions. These morphological changes interact with the more immediate effects of sea-level rise and will often exacerbate them.

Here only the most serious physical effects of sea-level rise are considered[56–58]

[44] R. J. T. Klein and R. J. Nicholls, Coastal zones, in J. F. Feenstra, I. Burton, J. B. Smith and R. S. J. Tol (eds.), *Handbook on Climate Change Impact Assessment and Adaptation Strategies*, Version 2.0, United Nations Environment Programme and Institute for Environmental Studies, Vrije Universiteit, Nairobi, Kenya, and Amsterdam, The Netherlands, 1998, pp. 7.1–7.35.

[45] M. Parry and T. Carter, *Climate Impact and Adaptation Assessment*, Earthscan, London, 1998.

[46] A. Sanchez-Arcilla, P. Hoekstra, J. Jimenez, E. Kaas and A. Maldonado, Climate change implications for coastal processes, in D. Smith, S. B. Raper, S. Zerbini and A. Sanchez-Arcilla (eds.), *Sea Level Change and Coastal Processes. Implications for Europe*, EUR 19337, European Commission, Brussels, 2000, pp. 173–213.

Table 2 The main effects of relative sea-level rise, including relevant interacting factors. Some factors (*e.g.* sediment supply) appear twice as they may be influenced both by climate and non-climate factors. A recent discussion of methods for assessment can be found in Klein and Nicholls[44]

Biogeophysical effect		*Other relevant factors*	
		Climate	*Non-climate*
Inundation, flood and storm damage[47–49]	Surge	Wave and storm climate, morphological change, sediment supply	Sediment supply, flood management, morphological change, land claim
	Backwater effect (river)	Run-off	Catchment management and land use
Wetland loss (and change)[47,49,50]		CO_2 fertilization Sediment supply	Sediment supply, migration space, direct destruction
Erosion[51–53]		Sediment supply, wave and storm climate	Sediment supply
Saltwater intrusion[54,55]	Surface waters	Run-off	Catchment management and land use
	Ground-water	Rainfall	Land use, aquifer use
Rising water tables/impeded drainage[54]		Rainfall	Land use, aquifer use

[47] F. M. J. Hoozemans, M. Marchand, M. and H. A. Pennekamp, *A Global Vulnerability Analysis: Vulnerability Assessment for Population, Coastal Wetlands and Rice Production on a Global Scale*, 2nd Edition, Delft Hydraulics, the Netherlands, 1993.

[48] F. M. J. Hoozemans and C. H. Hulsbergen, Sea-level rise: a world-wide assessment of risk and protection costs, in D. Eisma (ed.), *Climate Change: Impact on Coastal Habitation*, Lewis Publishers, London, 1995, pp. 137–163.

[49] R. J. Nicholls, F. M. J. Hoozemans and M. Marchand, Increasing flood risk and wetland losses due to global sea-level rise: regional and global analyses, *Global Environ. Change*, 1999, **9**, S69–S87.

[50] D. J. Reed, The response of coastal marshes to sea-level rise: survival or submergence?, *Earth Surf. Proc. Landforms*, 1995, **20**, 39–48.

[51] M.J.F. Stive, J.A. Roelvink and H.J. De Vriend, Large-scale coastal evolution concept, *Proceedings 22nd International Conference on Coastal Engineering*, ASCE, New York, 1990, pp. 1962–1974.

[52] R. J. Nicholls, Assessing erosion of sandy beaches due to sea-level rise, in J. G. Maund and M. Eddleston (eds.), *Geohazards in Engineering Geology*, Geological Society, London, Engineering Special Publication No. 15, 1998, pp. 71–76.

[53] P. J. Cowell, M. J. F. Stive, A. W. Niedoroda, H. J. de Vriend, D. J. P. Swift, G. M. Kaminsky and M. Capobianco, The Coastal-Tract (Part 1): a conceptual approach to aggregated modelling of low-order coastal change, *J. Coastal Res.*, in press.

[54] G. H. P. Oude Essink, *Impact of Sea level Rise on Groundwater Flow Regimes. A Sensitivity Analysis for the Netherlands.* PhD Thesis, Delft Technical University of Technology, The Netherlands, 1996, 411 pp.

[55] G. H. P. Oude Essink, Improving fresh groundwater supply – problems and solutions, *Ocean Coastal Manage.*, 2001, **44**, 421–441.

[56] M. C. Barth and J. G. Titus (eds.), *Greenhouse Effect and Sea Level Rise: A Challenge for this Generation.* Van Nostrand Reinhold, 1984, 325 pp.

[57] National Research Council, *Responding to Changes in Sea Level: Engineering Implications*, National Academy Press, Washington, DC, 1987.

[58] A. V. Tsyban, J. T. Everett and J. G. Titus, World ocean and coastal zones, in *Climate Change: The IPCC Impacts Assessment. Contribution of Working Group II to the First Assessment of the Intergovernmental Panel on Climate Change*, Australian Government Publishing Service, Canberra, Australia, 1990, pp. 86–93.

(Table 2). Naturally, low-lying coastal areas are most sensitive, including deltas, low-lying coastal plains, coral islands, beaches, barrier islands, coastal wetlands, and estuaries. The actual impacts will depend on human management, For instance, increased flooding and inundation of coastal lowlands will promote wetland creation, while protection of coastal lowlands will exacerbate wetland losses, as no replacement of losses will occur. Selected biogeophysical effects are now considered in more detail.

Inundation, floods and storm damage. Inundation (*i.e.* permanent submergence) of the Maldives and large parts of Bangladesh due to sea-level rise was one of the first forecasts that alerted the world community to the threat of human-induced climate change.[59,60] More generally, deltas and coral islands were recognized as being highly threatened by the effects of sea-level rise due to their low elevation. However, before an area is inundated, it will first experience an increasing frequency of flooding (*i.e.* temporary submergence) and storm damage as the existing flood plain is flooded more frequently, and the flood plain expands in size.[49,61] This effect will occur for both surges and river floods of coastal lowlands. Therefore, while most analyses distinguish inundation and increased flooding as distinct processes, they are part of a continuum.

Deltas form near sea level where river-deposited sediment accumulates in the coastal zone, and hence provide a good example of threatened coastal lowlands. Deltaic areas can often keep pace with rising sea level by riverborne sedimentation. However, delta management often excludes floods, while upstream dams reduce sedimentary inputs.[62,63] Hence this is an example of human management increasing vulnerability of coastal areas to sea-level rise. Sustaining sediment supply and developing flood management approaches that allow sedimentation are two approaches that could assist delta survival under sea-level rise based on a 'working with nature' philosophy.

Wetland loss (and change). Coastal wetlands (saltmarsh, mangroves and unvegetated intertidal mud) are sensitive to long-term sea-level change as their location is intimately linked to sea level. In response to sea-level rise, coastal wetlands experience faster vertical accretion due to increased sediment and

[59] J. M. Broadus, J. D. Milliman, S. F. Edwards, D. C. Aubrey and F. Gable, Rising sea level and damming of rivers: possible effects in Egypt and Bangladesh, in J. G. Titus (ed.), *Effects of Changes in Stratospheric Ozone and Global Change*, Volume 4, US Environmental Protection Agency, Washington DC, 1986, pp. 165–189.

[60] Commonwealth Secretariat, *Expert Group on Climate Change and Sea-Level Rise*. Commonwealth Secretariat, London, 1989, 230 pp.

[61] R. J. Nicholls, An analysis of the flood implications of the IPCC Second Assessment global sea-level rise scenarios, in D. J. Parker (ed.), *Floods*, Routledge, London, 2000, pp. 148–162.

[62] D. F. Boesch, M. N. Josselyn, A. J. Mehta, J. T. Morris, W. K. Nuttle, C. A. Simenstad and D. J. P. Swift, Scientific assessment of coastal wetland loss, restoration and management in Louisiana, *J. Coastal Res.*, 1994, Special Issue No. 20.

[63] A. Sanchez-Arcilla, J. Jimenez and H. I. Valdemoro, The Ebro delta: morphodynamics and vulnerability, *J. Coastal Res.*, 1998, **14**, 754–772.

organic matter input.[50,64,65] If vertical accretion equals sea-level rise, the coastal wetland will grow upwards in place. However, if accretion is less than sea-level rise, the coastal wetland steadily loses elevation relative to sea level. Vegetated wetlands are submerged for progressively longer periods during the tidal cycle and may die due to water-logging, causing a change to bare sediment, or even open water. Unvegetated intertidal areas are just progressively submerged. Therefore, coastal wetlands show a dynamic and non-linear response to sea-level rise.[49]

Direct losses of coastal wetland due to submergence (or edge erosion) can be offset by inland wetland migration (coastal dryland conversion to wetland). As sea level rises, so some low-lying coastal areas become suitable for the growth of wetland plants.[12] In areas without low-lying coastal areas, or in low-lying areas that are protected by humans to reduce coastal flooding, wetland migration cannot occur. This produces what is termed a 'coastal squeeze' between the defences and rising sea levels.[66] Therefore, adaptation to protect human use of the coastal zone may exacerbate wetland losses.

Coastal erosion. Bruun[67] suggested that there was a link between sea-level rise and shoreline recession on sandy beaches based on an equilibrium model of cross-shore beach response. On a typical beach, this might result in shoreline recession 100 times the rise in sea level. The Bruun Rule excites great passion and while there are several field studies that suggest it has some validity[68–70] the concept remains controversial.[71] Importantly, the Bruun Rule only describes one of many processes shaping sandy coasts. An additional erosional process linked to sea-level rise is the indirect effect of sea-level rise: as seas rise, estuaries and lagoons maintain equilibrium by raising their bed elevation in tandem, and act as a major sink for sand.[51,52,72] The sand is eroded from the open coast, potentially

[64] D. R. Cahoon, D. J. Reed and J. W. Day, Jr., Estimating shallow subsidence in microtidal saltmarshes of the southeastern United States: Kaye and Barghoorn revisited, *Marine Geol.*, 1995, **128**, 1–9.

[65] D. R. Cahoon and J. C. Lynch, Vertical accretion and shallow subsidence in a mangrove forest of southwestern Florida, U.S.A., *Mangroves Salt Marshes*, 1999, **1**, 173–186.

[66] R. J. Nicholls, Coastal zones, in M. L. Parry (ed.), *Assessment of the Potential Effects of Climate Change in Europe*, Jackson Environment Institute, University of East Anglia, 2000, pp. 243–259.

[67] P. Bruun, Sea level rise as a cause of shore erosion, *J. Waterways Harbors Div., ASCE*, 1962, **88**, 117–130.

[68] E. B. Hands, The Great Lakes as a test model for profile responses to sea level changes, in P. D. Komar (ed.), *Handbook of Coastal Processes and Erosion*, CRC Press, Boca Raton, FL, 1983, pp. 167–189.

[69] N. Mimura and H. Nobuoka, Verification of the Bruun Rule for the estimation of shoreline retreat caused by sea-level rise, *Proceedings of Coastal Dynamics 95*, Gdansk, Poland, 4–8 Sept. 1995, ASCE, New York, 1996, pp. 607–616.

[70] S. P. Leatherman, K. Zhang and B. C. Douglas, Sea level rise shown to drive coastal erosion, *EOS (Trans. Am. Geophys. Union)*, 2000, **81** (6), 55–57.

[71] P. D. Komar, *Beach and Nearshore Sedimentation*, Second Edition, Prentice Hall, Upper Saddle River, NJ, USA, 1998.

[72] M. A. Van Goor, M. J. F. Stive, Z. B. Wang and T. J. Zitman, Influence of relative sea level rise on coastal inlets and tidal basins, *Proceedings of Coastal Dynamics 2001*, ASCE, New York, 2001, pp. 242–251.

causing ten times the erosion predicted by the Bruun Rule in the vicinity of inlets.

Bird[73] estimated that 70% of the world's sandy beaches are already eroding. Bird offered several alternative explanations, including widespread sediment starvation and global sea-level rise. Whatever the present cause, it is widely agreed that an acceleration in sea-level rise will promote erosion.[12] New models describing long-term coastal behaviour offer promise to improve prediction of future coastal evolution, integrating all the forces shaping the coast including sea-level rise.[53,74,75]

Socio-Economic Impacts of Sea-level Rise

The range of natural-system effects of sea-level rise in Table 2 have a wide range of potential socio-economic impacts,[43] reflecting the diverse uses of the coastal zone. The IPCC Common Methodology defined three types of impacts:[42,47]

- *values at risk* due to instantaneous changes such as increased flood frequency and depth, and storm damage for people, land and infrastructure;
- *values at loss* due to longer-term effects such as land loss and abandonment in response to erosion and inundation, and ecosystem loss. This could include direct loss of economic, ecological, cultural and subsistence values through loss of land, infrastructure, water resources and coastal habitats;
- *values at change* where the consequences of sea-level rise are uncertain due to incomplete knowledge and/or there may be both positive and negative effects, such as some hydrological changes. Further analysis will lead to values at change being reclassified to the other two categories, or no impacts.

Table 3 lists the most important socio-economic sectors in coastal zones, and indicates which natural system effects are expected to cause direct socio-economic impacts, including the cases where it is uncertain. Indirect impacts of sea-level rise are not shown as they are more difficult to analyse, but they have the potential to be important in many sectors, such as human health.[76] Examples of possible triggers of indirect health impacts include the nutritional impacts of loss of agricultural production in coastal areas, the release of toxic materials from abandoned land fills during storms,[77,78] and possible effects due to waterlogging

[73] E. C. F. Bird, *Coastline Changes: A Global Review*, John Wiley, Chichester, 1985, 219 pp.

[74] A. W. Niedoroda, C. W. Reed, M. J. F. Stive and P. Cowell, P., Numerical simulations of coastal-tract morphodynamics, *Proceedings of Coastal Dynamics 2001*, ASCE, New York, 2001, pp. 403–412.

[75] M. J. F. Stive, S. J. C. Aarninkoff, L. Hamm, H. Hanson, M. Larson, K. Wijnberg, R. J. Nicholls and M. Capobianco, Variability of shore and shoreline evolution, *Coastal Eng.*, accepted.

[76] A. McMichael, A. Githeko, R. Akhtar, R. Carcavallo, D. Gubler, A. Haines, S. Kovats, P. Martens, J. Patz and A. Sasaki, Human health, in J. J. McCarthy, O. F. Canziani, N. A. Leary, D. J. Dokken and K. S. White (eds.), *Climate Change 2001: Impacts, Adaptation and Vulnerability*, Cambridge University Press, Cambridge, 2001, pp. 451–485.

[77] T. J. Flynn, S. G. Walesh, J. G. Titus and M. C. Barth, Implications of sea level rise for hazardous waste sites in coastal floodplains, in M. C. Barth and J. G. Titus (eds.), *Greenhouse Effect and Sea Level Rise: A Challenge for this Generation*, Van Nostrand Reinhold, 1984, pp. 271–294.

[78] J. E. Neumann, G. Yohe, R. J. Nicholls and M. Manion, *Sea Level Rise and Global Climate Change: A Review of Impacts to US Coastal Resources*. Climate Change Briefing Paper of the Pew Foundation, Washington DC, 2000, 38 pp.

Table 3 Qualitative analysis indicating the more significant *direct* socio-economic impacts of climate change and sea-level rise on some of the different sectors in coastal zones, including coastal biodiversity

Sector	*Biogeophysical effect (from Table 2)*						
	Inundation, flood and storm damage				*Saltwater intrusion*		
	Surge	*Backwater effect*	*Wetland loss*	*Erosion*	*Surface*	*Ground*	*Rising water tables*
Water Resources	✓				✓	✓	
Agriculture	✓	✓			✓	✓	✓
Human health	✓	✓			✓	✓	
Fisheries	?	?	✓		✓		
Tourism	✓		✓	✓			
Human settlements	✓	✓		✓	✓	✓	✓
Coastal biodiversity	?	?	✓	✓	✓	✓	✓

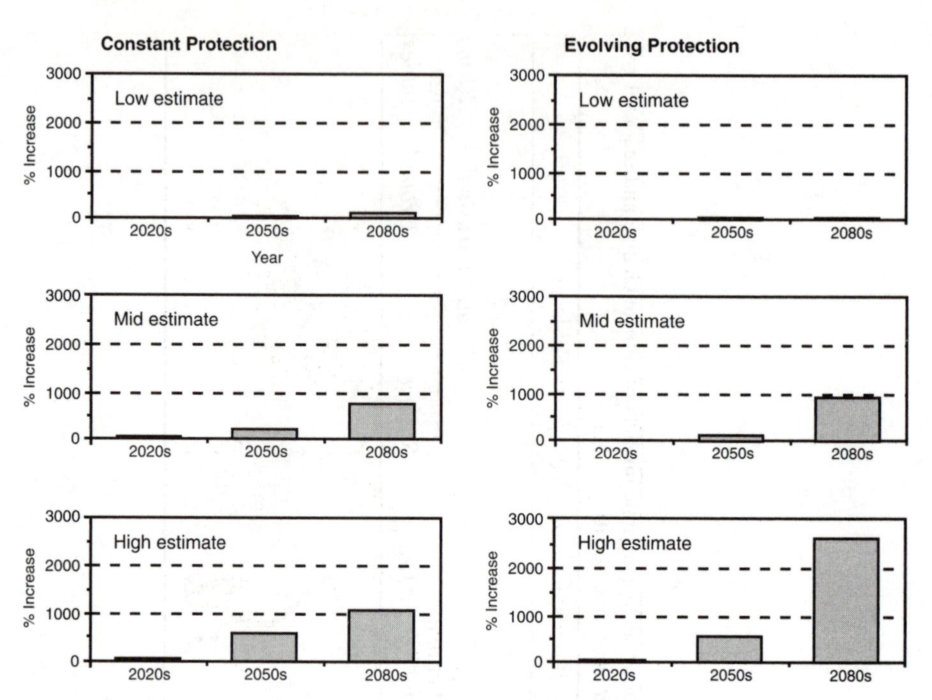

Figure 5 The flood impacts of the IPCC IS92a scenarios (taken from Nicholls[61]). Constant protection assumes population growth and constant 1990 defences, while evolving protection assumes population growth and upgraded defence standards as national wealth increases, but with no allowance for sea-level rise Reproduced by permission of Routledge

and rising water tables. Thus, sea-level rise can produce a cascade of impacts through the socio-economic system.

Selected Regional and Global Impact Results

To illustrate the possible impacts of sea-level rise, selected regional and global results are now considered. Globally, about 200 million people lived in the coastal flood plain (below the 1 in 1000 year surge-flood elevation) in 1990, or about 4% of the world's population.[47] It is estimated that on average 10 million people/year experience flooding.[49] Even without sea-level rise, this number will increase significantly due to increasing coastal populations. Figure 5 shows the relative impacts of the IS92a global-mean sea-level rise scenarios (a rise in the range 19–80 cm from 1990 to the 2080s). The relative increase in the number of people flooded is relatively minor under the low scenario and up to 25 times under the high scenario with > 350 million people experiencing flooding each year.[61] In the latter case, most of these people will be flooded so frequently (more than once per year) that a response seems inevitable (migration, upgraded defences, *etc.*). The most vulnerable regions in relative terms are the island regions of the Caribbean, Indian Ocean and Pacific Ocean small islands. Absolute increases are largest in the southern Mediterranean, West Africa, East Africa, South Asia and South-East Asia (Figure 6).

Upgrading coastal protection infrastructure against a 1 m rise in sea level is estimated to have a global cost of US $1000 billion (1990 dollars), or 5.6% of the 1990 Global World Income.[47,48] These cost estimates assume an instantaneous rather than a progressive response and do not consider erosion in non-tourist

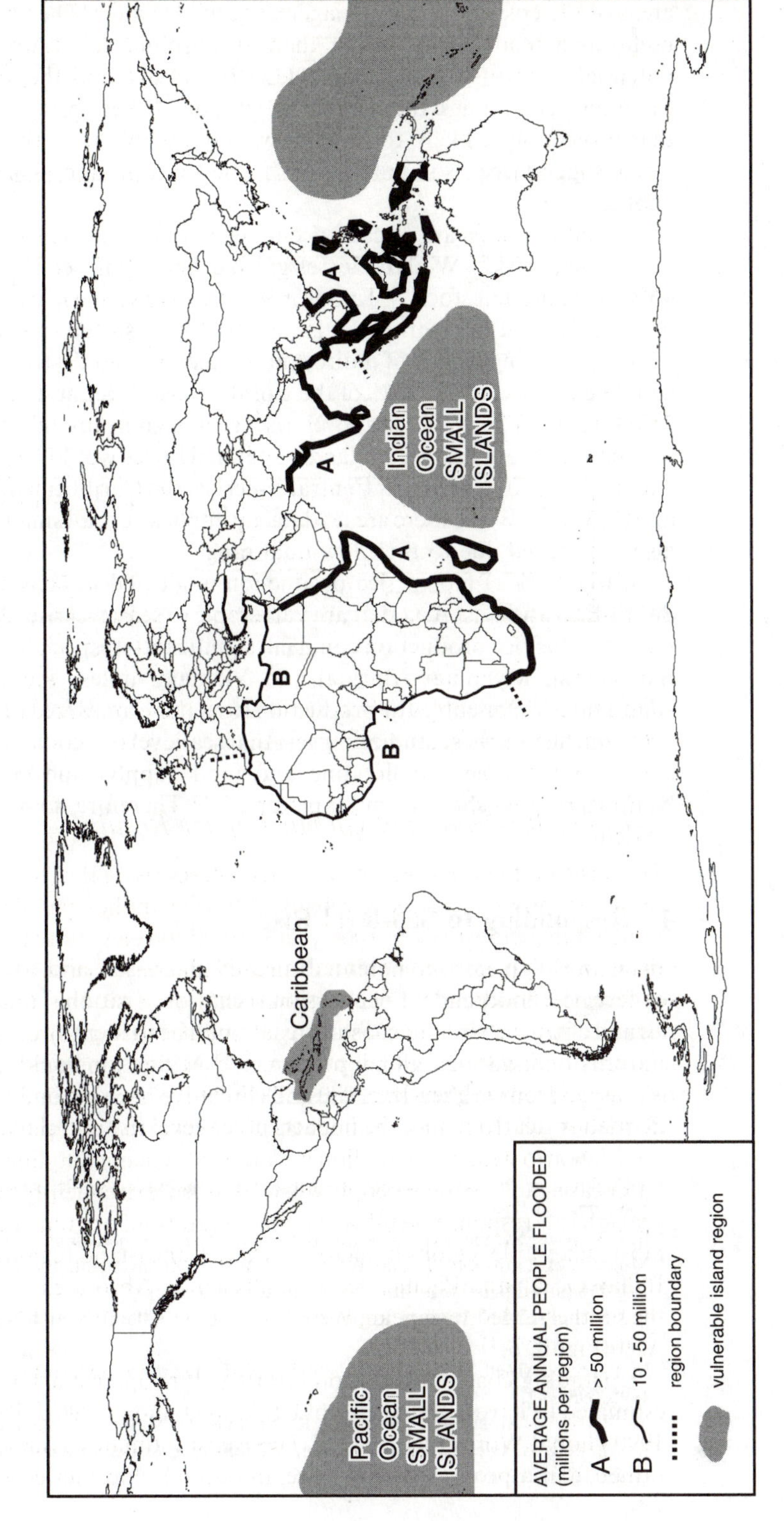

Figure 6 The regions most affected by flood impacts in the 2080s given the high estimate sea-level rise scenario and evolving protection (taken from Nicholls[61])
Reproduced by permission of Routledge

areas or the costs of water management and drainage. Therefore, they are more useful as a relative cost rather than an absolute adaptation cost. Given the potentially serious flood impacts described above and the long-term trend of expanding economies, the relative magnitude of these protection costs suggests that protection is likely to be more widely feasible than the current situation would suggest (see also Fankhauser[79]). The appropriateness of such protection is another question.

Coastal wetlands are already declining at 1%/year due to indirect and direct human activities.[47] Wetland losses given a 1 m rise in sea level could approach 46% of the present stock.[49] Taking a 38 cm global scenario by the 2080s, between 6% and 22% of the world's wetlands could be lost due to sea-level rise. When added to existing trends of indirect and direct human destruction, the net effect could be the loss of 36–70% of the world's coastal wetlands, or an area of up to 210 000 km^2. Therefore, sea-level rise is a significant additional stress that worsens the prognosis for wetlands. Regional losses would be most severe on the Atlantic coast of North and Central America, the Caribbean, the Mediterranean and the Baltic. While there are no data, by implication, all small island regions are also threatened due to their low tidal range.

Nearly 10% of global rice production is located in coastal areas in South, South-East and East Asia that are vulnerable to sea-level rise. Based on a 1 m rise scenario, this rice production could fall significantly, especially the large deltas of Bangladesh, Myanmar (Burma) and Vietnam, unless there was substantial adaptation.[47] Presently, the production from these areas feeds 200 million people.

In conclusion, these studies suggest that sea-level rise could produce important impacts on people (*via* flooding and food supply) and coastal ecosystems. National reviews show a similar picture.[80,81] Therefore, some response appears prudent.

4 Responding to Sea-level Rise

Given the high impact potential already discussed and the commitment to sea-level rise independent of emission scenarios, a rational response to sea-level rise and climate change in coastal areas would be to identify the most appropriate mixture of mitigation and adaptation.[82,83] Mitigation could greatly reduce the risks associated with sea-level rise and climate change beyond the 21st century.[14] Adaptation acts to reduce the impacts of sea-level rise and climate change, as well

[79] S. Fankhauser, Protection versus retreat: estimating the costs of sea-level rise, *Environ. Planning A*, 1995, **27**, 299–319.

[80] R. J. Nicholls, Synthesis of vulnerability analysis studies, *Proceedings of WORLD COAST 1993*, Ministry of Transport, Public Works and Water Management, the Netherlands, 1995, pp. 181–216 (downloadable at http//www.survas.mdx.ac.uk/).

[81] R. J. Nicholls and N. Mimura, Regional issues raised by sea-level rise and their policy implications, *Clim. Res.*, 1998, **11** (1), 5–18.

[82] M. Parry, N. Arnell, M. Hulme, R. Nicholls and M. Livermore, Adapting to the inevitable, *Nature*, 1998, **395**, 741.

[83] M. Parry, N. Arnell, T. McMichael, R. Nicholls, P. Martens, S. Kovats, M. Livermore, C. Rosenzweig, A. Iglesias and G. Fischer, Millions at risk: defining critical climate threats and targets, *Global Environ. Change*, in press.

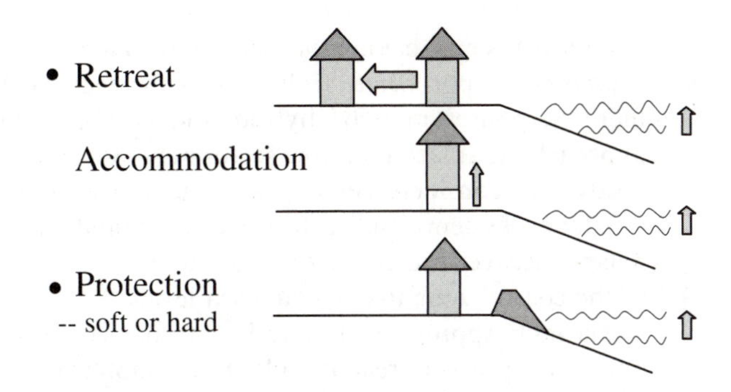

Figure 7 The three generic response options to sea-level rise[85]

as other changes (as well as exploiting benefits). These decisions need to be made in the face of the large uncertainty about future climate (and many other factors), so there is a need to think in a risk- and uncertainty-based manner rather than looking for deterministic solutions.

Given the large and growing concentration of people and activity in the coastal zone, autonomous adaptation processes are unlikely to be sufficient to respond to sea-level rise[44] (although the distinction between autonomous and planned adaptation is not always clear). Further, adaptation in the coastal context is widely seen as a public responsibility.[84] Therefore, all levels of government have a key role in developing and facilitating appropriate adaptation measures. This will probably include policies to stimulate autonomous processes of adaptation, which is one way to enhance adaptive capacity (and is often reversing historic trends). Planned adaptation options to sea-level rise are usually presented as one of three generic approaches[12,85] (Figure 7):

- *(Planned) Retreat* – all natural system effects are allowed to occur and human impacts are minimized by pulling back from the coast;
- *Accommodation* – all natural system effects are allowed to occur and human impacts are minimized by adjusting human use of the coastal zone;
- *Protection* – natural system effects are controlled by soft or hard engineering, reducing human impacts in the zone that would be impacted without protection.

Through human history, improving technology has increased the range of adaptation options in the face of coastal hazards, and there has been a move from retreat and accommodation to hard protection and active seaward advance *via* land claim. Rising sea level is one factor calling universal reliance on hard protection into question, and the appropriate mixture of protection, accommodation

[84] R. J. T. Klein, J. Aston, E. N. Buckley, M. Capobianco, N. Mizutani, R. J. Nicholls, P. D. Nunn and S. Ragoonaden, Coastal-adaptation technologies, in B. Metz, O. R. Davidson, J. W. Martens, S. N. M. van Rooijen and L. L. Van Wie McGrory (eds.), *IPCC Special Report on Methodological and Technological Issues in Technology Transfer*, Cambridge University Press, Cambridge, 2000, pp. 349–372.

[85] IPCC CZMS, *Strategies for Adaptation to Sea Level Rise*, Report of the Coastal Zone Management Subgroup, Response Strategies Working Group of the Intergovernmental Panel on Climate Change, Ministry of Transport, Public Works and Water Management, The Hague, The Netherlands, 1990, x + 122 pp.

and retreat is now being more seriously evaluated.[86,87] Klein *et al.*[88] have given examples of appropriate technologies for each of these measures. In practice, many responses may be hybrid and combine elements of more than one approach. Adaptation for one sector may exacerbate impacts elsewhere and this needs to be considered in any assessment. A good example is coastal squeeze of coastal ecosystems due to hard defences, and there is a need to consider the balance between protecting socio-economic activity and the ecological functioning of the coastal zone under rising sea levels.[66]

The most appropriate timing for a response needs to be considered in terms of anticipatory *versus* reactive planned adaptation (or in practical terms, what should we do today *versus* wait and see until tomorrow?). Anticipatory decisions are made with more uncertainty than reactive decisions, which will have the benefit of future knowledge. However, wait and see may lock in an adverse direction of development which increases exposure to sea-level rise. In terms of impacts, Figure 8 summarizes these decisions. Sea-level rise has potential impacts. Anticipatory planned adaptation can reduce these potential impacts to the initial impacts. Reactive adaptation (including autonomous adaptation) in response to the initial impacts further reduces the impacts to the residual impacts. The realistic magnitude of the initial and residual impacts is a key measure of vulnerability, although many assessments focus only on evaluating potential impacts.

The coastal zone is an area where anticipatory adaptation needs to be carefully considered as many decisions at the coast have long-term implications.[90,91] Examples of anticipatory adaptation in coastal zones include upgraded flood defences and waste water discharges, higher levels for reclamation and new bridges, and building setbacks to prevent development.[11,12]

While there is limited experience of adaptation to climate change, there is considerable experience of adapting to climate variability and we can draw on this experience to inform decision making under a changing climate.[92] An analysis of the evolution of coastal zone management in the Netherlands, UK and Japan shows that adaptation to coastal problems is a process, rather than just the implementation of technical options. Four stages in the adaptation process related to (1) information and awareness building, (2) planning and design, (3) evaluation and (4) monitoring and evaluation were evident within

[86] J. H. M. De Ruig, Coastline management in the Netherlands: human use versus natural dynamics, *J. Coastal Conserv.*, 1998, **4**, 127–134.

[87] R. J. T. Klein, M. J. Smit, H. Goosen and C. H. Hulsbergen, Resilience and vulnerability: coastal dynamics or Dutch dikes?, *Geogr. J.*, 1998, **164**, 259–268.

[88] R. J. T. Klein, R. J. Nicholls, S. Ragoonaden, M. Capobianco, J. Aston and E. N. Buckley, Technological options for adaptation to climate change in coastal zones, *J. Coastal Res.*, 2001, **17**, 531–543.

[89] R. J. T. Klein, Towards better understanding, assessment and funding of climate adaptation, *Change*, 1998, **44**, 15–19.

[90] J. B. Smith, Setting priorities for adapting to climate change, *Global Environ. Change*, 1997, **7**, 251–264.

[91] S. Fankhauser, J. B. Smith and R. S. J. Tol, Weathering climate change: some simple rules to guide adaptation decisions, *Ecol. Econ.*, 1999, **30**, 67–78.

[92] R. J. T. Klein, R. J. Nicholls and N. Mimura, Coastal adaptation to climate change: can the IPCC Technical Guidelines be applied?, *Mitigation and Adaptation Strategies for Global Change*, 1999, **4**, 51–64.

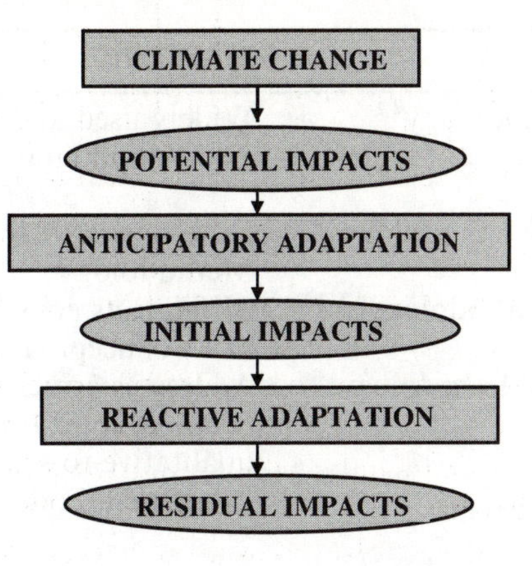

Figure 8 Framework for impact and adaptation assessment (adapted from Klein[89])

multiple policy cycles, as well as various constraints on approaches to adaptation due to broader policy and development goals. Green *et al.*[93] argue along consistent lines that an adaptive management approach to climate change is clearly appropriate given the entire climate and other uncertainties (see also National Research Council[94]). An explicit 'learning by doing' approach accepts uncertainty and sees any intervention in the environment as an educated experiment and an opportunity to learn *via* careful monitoring and evaluation. It also recognizes that the criteria for assessing the suitability of an adaptation approach are likely to evolve due to improving scientific and technical knowledge and/or changing societal values.

With a few exceptions, climate change will largely exacerbate existing pressures and problems, so there are important synergies in considering adaptation to climate change in the context of existing problems.[95,96] In some cases, the focus of sea-level rise and climate change may help identify 'win-win' situations that are worthy of implementation without any climate change. Also, adaptation measures are more likely to be implemented if they offer immediate benefits in reducing impacts of short-term climate variability as well as long-term climate change.

Last, broader measures in terms of enhancing adaptive capacity and creating an environment where adaptation can more easily occur are also vital (*cf.* Smit *et al.*[41]). In many countries there is limited capacity to address today's coastal problems, let alone consider tomorrow's problems, including sea-level rise. Therefore, such capacity building should include developing coastal management, as already widely recommended.[4,85,97]

[93] C. Green, R. J. Nicholls, C. Johnson, S. Shackley and J. W. Handmer, *Climate Change Adaptation: A Framework for Analysis and Decision-making in the Face of Risk and Uncertainty*, Report No. 28, A report produced for NCRAOA as part of Environment Agency R&D project E2-036, Environment Agency, 2000, Bristol, 87 pp.

[94] National Research Council, *Science, Policy and the Coast: Improving Decisionmaking*, National Academy Press, Washington, DC, 1995, 85 pp.

[95] R. A. Pielke, Rethinking the role of adaptation in climate policy, *Global Environ. Change*, 1998, **8**, 159–170.

[96] R. J. T. Klein, Adaptation to climate change: what is optimal and appropriate?, *Clim. Change*, 2001, in press.

Table 4 Different methods for vulnerability assessment in coastal areas

Methods	Comments
IPCC Common Methodology[42]	Widely used and criticized[11,44]
EPA International Project Methodology[98]	Focuses on land loss impacts
US Country Studies Methodology[99]	Develops EPA International Project Methodology
UNEP Handbook Methodology[43,44]	All impacts considered. Uses Figure 4 as its conceptual base
South Pacific Islands Methodology[100]	Addresses criticisms of Common Methodology, but results are qualitative to semi-quantitative
RIKS Decision Support Methodology[101]	Integrated model approach

5 The Future of Vulnerability Assessments

Sea-level rise and climate change presents an important challenge for the future, and yet our understanding of the issues, particularly with regard to adaptation, remains poor. A range of vulnerability assessment methodologies are available, although many are related (Table 4). The UNEP Guidelines[44] were developed using the conceptual model shown in Figure 4. To further help structure future assessments, Table 5 suggests a hierarchy of assessments from 'quick and dirty' screening assessments, through traditional vulnerability assessments, to studies that are designed to provide coastal planners and managers with detailed technical guidance. With this suggested structure in mind, it is hoped that factors such as effort, amount of data collection and appropriate models could be made more explicit. Considering erosion and related changes, a screening assessment might use the Bruun 'rule-of-thumb', a vulnerability assessment might use the Bruun rule coupled to a longshore transport model, while a planning assessment might use a suite of models within a GIS framework and bounded using the coastal tract cascade.[52,53,102]

[97] B. Cicin-Sain, C. Ehler, R. Knecht, R. South and R. Weiher, *Guidelines for Integrating Coastal Management Programs and National Climate Change Action Plans*, International Workshop: Planning for Climate Change Through Integrated Coastal Management, February 1997, Taipei. NOAA, Silver Spring, MD, 1997.

[98] R.J. Nicholls and S.P. Leatherman (eds.), Potential impacts of accelerated sea-level rise on developing countries, *J. Coastal Res.*, 1995, Special Issue No. 14.

[99] S.P. Leatherman and G.W. Yohe, Coastal impact and adaptation assessment, in R. Benioff, S. Guill and J. Lee (eds), *Vulnerability and Adaptation Assessments – An International Handbook*, Version 1.1, Kluwer Academic Publishers, Dordrecht, 1996, pp. 5.63–5.76, H.1–H.39.

[100] K. Yamada, P.D. Nunn, N. Mimura, S. Machida and M. Yamamoto, Methodology for the assessment of vulnerability of South Pacific island countries to sea-level rise and climate change. *J. Global Environ. Eng.*, 1995, **1**, 101–125.

[101] G. Engelen, R. White, I. Uljee and S. Wargnies, Numerical modeling of small island socio-economics to achieve sustainable development, in G.A. Maul (ed.), *Small Islands: Marine Science and Sustainable Development*, Coastal and Estuarine Studies Volume 51, American Geophysical Union, Washington DC, 1996, p. 437–463.

[102] M. Capobianco, H.J. DeVriend, R.J. Nicholls and M.J.F. Stive, Coastal area impact and vulnerability assessment: a morphodynamic modeller's point of view, *J. Coastal Res.*, 1999, **15**, 701–716.

Table 5 Three levels of assessment in coastal zones, showing the respective requirements and the factors to be considered (taken from Klein and Nicholls;[43] see also Nicholls[52])
Reproduced with permission of the Royal Academy of Sciences

Level of assessment	*Requirements*			*Factors to consider*		
	Time	*Level of detail*	*Socio- Prior knowledge*	*Other economic factors*	*Non-climatic changes*	*climate changes*
Screening assessment	2–3 months	Low	Low	No	No	No
Vulnerability assessment	6–12 months	Medium	Medium	Yes	Possible	No
Planning assessment	Continuous	High	High	Yes	Yes	Yes

Therefore, it is important to continue to develop both the science of vulnerability assessment and its application to inform coastal policy at all levels. By definition, this must be a multi-disciplinary activity which brings together relevant natural, engineering and social sciences in a policy-relevant manner. Review of the available studies and experience shows that there are still important barriers to conducting comprehensive impact and response analyses,[103] including:

- incomplete knowledge of the relevant processes affected by sea-level rise and their interactions;
- insufficient data on existing conditions;
- difficulty in developing the local and regional scenarios of future change;
- lack of appropriate analytical methodologies for some impacts.

Most existing vulnerability assessments have focussed on a 1 m rise in global sea level on the present (1990s) situation applied directly as a relative sea-level rise scenario.[80,81] In addition, only the direct effects of sea-level rise such as inundation and erosion were considered and the dynamic processes that produce autonomous adaptation such as wetland response to sea-level rise were ignored (*i.e.* sea-level rise is instantaneous, rather than progressive). If they are interpreted with the relevant assumptions in mind, these studies provide important insights into the potential impacts of sea-level rise. However, if taken literally, they are clearly unrealistic.

It is noteworthy that many studies have difficulties defining appropriate climate scenarios and spend a disproportionate amount of effort on this stage. Given the large uncertainties about future relative sea levels, particularly when meteo-oceanographic effects are considered, it would be prudent for future vulnerability assessments to consider side-stepping this issue by developing response surfaces for sea-level rise scenarios from no rise up to 1.5 m by 2100. Therefore, as scientific knowledge on regional sea-level rise improves, so the impacts can be rapidly reinterpreted based on existing studies. For long-term issues (*e.g.* nuclear power plants in coastal locations), it would also be prudent to consider the potential impacts of collapse of the WAIS, at least at the qualitative level.

Most recently, the SURVAS Project considered the future of vulnerability assessment studies. It was recommended that future assessments consider the following issues.

- Place sea-level rise in a broader context of change and today's problems including consideration of:
  - other climate change;
  - extreme events;
  - non-climate changes;
- Assess the cascade of impacts from the natural system to and through the

[103] R. J. Nicholls and A. C. de la Vega-Leinert, Introduction/Recommendations, *Proceedings of the SURVAS Overview Workshop on the Future of Vulnerability and Adaptation Studies*, 28–30 June 2001, London, UK, Flood Hazard Research Centre, Middlesex University, UK, in preparation (will be available online at: http://www.survas.mdx.ac.uk).

socio-economic system (Figure 4), including direct and indirect impacts;
- Consider sub-tidal and inter-tidal changes and impacts (mainly fisheries and ecosystems), as well as terrestrial changes;
- Identify 'flagship impacts' on cultural or natural sites (*e.g.* Venice, Italy or the Carmague, France) that are likely to attract public attention to the issue;
- Identify any adaptation measures that should be implemented immediately and consider the barriers and constraints to adaptation;
- Last, consider the capacity to adapt, which is related to adaptive capacity.

6 Discussion/Conclusions

Human-induced sea-level rise and other climate change may adversely affect coastal zones around much of the world's coasts in the 21st century and beyond. Reducing greenhouse gas emissions (mitigation) only reduces the magnitude of these changes and an appropriate combination of mitigation and adaptation appears to be the best policy response to climate change in coastal areas.

To date, few studies have assessed the impacts of sea-level rise in a realistic manner, including the potential to adapt. Evaluating adaptation requires a full understanding of the potential impacts and the adaptive capacity. The latter factor embraces autonomous adaptation, and (anticipatory and reactive) planned adaptations. This allows an assessment of the impacts after likely adaptation, and hence a more realistic view of the vulnerability to sea-level rise. Such analysis requires a conceptual model of the coastal zone, including recognizing the interacting natural and socio-economic systems, definition of the different impacts that might be assessed, recognition of the adaptation process, and a consistent vulnerability assessment methodology. Given the role of other non-climate stresses on the impacts of sea-level rise (and climate change), it is also important that these are considered. The full range of model tools can only be meaningfully integrated and brought to bear on the problem within this wider conceptual framework.

While all aspects of vulnerability assessment require further research, the aspects of integration and adaptation require the most development. The concept of adaptive capacity has been proposed, and the challenge is to make it meaningful in the coastal context such that 'enhancing adaptive capacity' could become an operational reality.

Acknowledgements

This article benefits from many collaborations over the years, particularly with Richard Klein (PIK) and Anne de la Vega-Leinert (PIK), and all the contributors to the SURVAS Project (http://www.survas.mdx.ac.uk/). The author would like to acknowledge a Leverhulme Trust Fellowship, which allowed this article to be prepared.

Climate Change, Global Food Supply and Risk of Hunger

MARTIN PARRY AND MATTHEW LIVERMORE

1 Introduction

In this article we review a number of studies of the possible effects of climate change on global agricultural yield potential, on cereal production, food prices and the implications for changes in the number of hungry people. At present, almost 800 million people in the developing world are estimated to be to experiencing some form of shortage in food supply.[1] In general, the conclusion from recent research has been that, while one may be reasonably optimistic about the prospects of adapting the agricultural production system to the early stages of global warming, the distribution of the vulnerability among the regions and people is likely to be uneven. Where crops are near their maximum temperature tolerance and where dryland, non-irrigated agriculture predominates, yields are likely to decrease with even small amounts of climate change. The livelihoods of subsistence farmers and pastoral people, who are already weakly coupled to markets, could also be negatively affected. In regions where there is a likelihood of decreased rainfall, agriculture could be substantially affected regardless of latitude.[2]

Clearly, in addition to the above generality on productivity, other features of agricultural vulnerability are likely to vary widely among people, regions, nations and continents. The poor, especially those living in marginal environments, will be most vulnerable to climate-induced food insecurity. By the 2080s, the additional number of people at risk of hunger due to climate change is estimated to be about 80 million. However, some regions (particularly in the arid and sub-humid tropics) may be affected more. Africa is projected to experience marked reductions in yield, decreases in production and increases in the risk of hunger as a result of climate change. Recent studies[3,4] suggest that the continent

[1] FAO, 1999. *The State of Food Insecurity in the World*, United Nations Food and Agriculture Organization, Rome, 1999.

[2] IPCC, *Climate Change 2001: Impacts, Adaptation and Vulnerability – Contribution of Working Group II to the IPCC Third Assessment*, Cambridge University Press, Cambridge, 2001.

[3] M. L. Parry, C. Rosenzweig, A. Iglesias, G. Fischer and M. T. J. Livermore, The impact of climate change on food supply, in *Climate Change and Its Impacts: A Global Perspective*, Department of the Environment, Transport and the Regions/UK Meteorological Office, 1997, pp. 12–13.

Issues in Environmental Science and Technology, No. 17
Global Environmental Change

can expect to have between 55 and 65 million extra people at risk of hunger by the 2080s.

The story of how we have arrived at this conclusion is traced below. We outline the research method, its testing and the first evaluation of effects on food supply with those climate change scenarios available in the early 1990s. This is followed by an assessment of effects under some of the most recent scenarios developed for the Intergovernmental Panel on Climate Change (IPCC).[5] Finally, we consider the effects of various mitigation strategies, involving stabilization of CO_2 concentrations, on reducing the scale of impact.

2 Initial Estimations for Climate Scenarios from Low Resolution Climate Models

The first studies of effects on global food supply, at least those that were model-based, were published in the early 1990s and were dependent at that time on climate scenarios that were the product of low resolution global climate models, before there was an effectively modelled coupling of ocean–atmosphere interactions. But the general conclusions of that work still hold today: that climate change is likely to reduce global food potential and that risk of hunger will increase in the most marginalized economies.[6] In the study two main components were considered:

1. The estimation of potential changes in crop yield. Potential changes in national grain crop yields were estimated using crop models and a decision support system developed by the US Agency for International Development's International Benchmark Sites Network for Agrotechnology Transfer (IBSNAT).[7] The crops modelled were wheat, rice, maize and soybean. These crops account for more than 85% of the world's traded grains and legumes. The estimated yield changes for 18 countries were interpolated to provide estimates of yield changes for all regions of the world and for all major crops, by reference to all available published and unpublished information.
2. Estimation of world food trade responses. The yield changes were used as inputs into a world food trade model, The Basic Linked System (BLS) developed at the International Institute for Applied Systems Analysis

[4] M. L. Parry, C. Rosenzweig, A. Iglesias, G. Fischer and M. T. J. Livermore, Climate change and world food security: a new assessment, in M. L. Parry and M. T. J. Livermore (eds.), A new assessment of the global effects of climate change, *Global Environ. Change*, 1999, **9** (Supplemental Issue), s52–s67.

[5] IPCC, R. T. Watson, M. C. Zinyowera and R. H. Moss (eds.), *Climate Change 1995: Impacts, Adaptations and Mitigation of Climate Change: Scientific-Technical Analyses – Contribution of Working Group II to the Second Assessment Report of the Intergovernmental Panel on Climate Change*, Cambridge University Press, Cambridge, 1995.

[6] C. Rosenzweig and M. L. Parry, Potential impacts of climate change on world food supply. *Nature*, 1994, **367**, 133–138.

[7] International Benchmark Sites Network for Agrotechnology Transfer (IBSNAT), *Decision Support System for Agrotechnology Transfer Version 2.1 (DSSAT V2.1)*. Dept. Agron. & Soil Sci. College Trop. Agric. And Hum. Resources. Univ. Hawaii, Honolulu, 1989.

(IIASA).[8] Outputs from simulations by the BLS provided information on food production, food prices and the number of people at risk of hunger.

Climate Change Scenarios

Scenarios of climate change were developed to estimate the effect on yields and food supply. The range of scenarios used aimed to capture the range of possible effects and set limits on the associated uncertainty. The scenarios for this study were created by changing the observed data on current climate (1951–80) according to doubled CO_2 simulations of three general circulation models (GCMs). The GCMs used were those from the Goddard Institute for Space Studies (GISS),[9,10] Geophysical Fluid Dynamics Laboratory (GFDL)[11] and the United Kingdom Meteorological Office (UKMO).[12]

Crop Models and Yield Simulations

Crop models. The IBSNAT crop models were used to estimate how climate change and increasing levels of carbon dioxide may alter yields of work crops at 112 sites in 18 countries representing both major production areas and vulnerable regions at low, mid and high latitudes.[13] The IBSNAT models employ simplified functions to predict the growth of crops as influenced by the major factors that affect yields, *e.g.* genetics, climate (daily solar radiation, maximum and minimum temperatures and precipitation), soils and management practices. Models used were for wheat,[14,15] maize,[16,17] paddy and upland rice[18] and soybean.[19]

[8] G. Fischer, K. Frohberg, M.A. Keyzer and K.S. Parikh, *Linked National Models: A Tool for International Food Policy Analysis*, Kluwer, Dordrecht, 1988.

[9] J. Hansen, G. Russell, D. Rind, P. Stone, A. Lacis, S. Lebedeff, R. Ruedy and L. Travis, Efficient three-dimensional global models for climate studies: Models I and II, *Monthly Weather Rev.*, 1983, **111**, 609–662.

[10] J. Hansen, I. Fung, A. Lacis, D. Rind, G. Russell, S. Lebedeff, R. Ruedy and P. Stone, Global climate changes as forecast by the GISS 3-D model, *J. Geophys. Res.*, 1988, **93**, 9341–9364.

[11] S. Manabe and R.T. Wetherald, Large-scale changes in soil wetness induced by an increase in CO_2, *J. Atmos. Sci.*, 1987, **44**, 1211–1235.

[12] C.A. Wilson and J.F.B. Mitchell, A doubled CO_2 climate sensitivity experiment with a global climate model including a simple ocean, *J. Geophys. Res.*, 1987, **92** (13), 315–343.

[13] C. Rosenzweig and A. Iglesias, *Implications of Climate Change for International Agriculture: Crop Modeling Study*, US Environmental Protection Agency, Washington, DC, 1994.

[14] J.T. Ritchie and S. Otter, Description and performance of CERES-Wheat: A user-oriented wheat yield model, in W.O. Willis (ed.), *ARS Wheat Yield Project*, Department of Agriculture, Agricultural Research Service, Washington, DC, ARS-38, 1985.

[15] D. Godwin, J.T. Ritchie, I.J. Singh and L. Hunt, *A User's Guide to CERES-Wheat – V2.10*, Muscle Shoals: International Fertilizer Development Center, 1989.

[16] C.A. Jones and J.R. Kiniry, *CERES-Maize: A Simulation Model of Maize Growth and Development*, Texas A&M Press, College Station, 1986.

[17] J.T. Ritchie, U. Singh, D. Godwin and L. Hunt, *A User's Guide to CERES-Maize–V2.10*, Muscle Shoals: International Fertilizer Development Center, 1989.

[18] D. Godwin, U. Singh, J.T. Ritchie and E.C. Alocilja, *A User's Guide to CERES-Rice*, Muscle Shoals: International Fertilizer Development Center, 1993.

[19] J.W. Jones, K.J. Boote, G. Hoogenboom, S.S. Jagtap and G.G. Wilkerson, *SOYGRO V5.42: Soybean Crop Growth Simulation Model Users' Guide*, Department of Agricultural Engineering and Department of Agronomy, University of Florida, Gainesville, 1989.

The IBSNAT models were selected for this study because they have been validated over a wide range of environments[20] and are not specific to any particular location or soil type. They are thus suitable for use in large-area studies in which crop growing conditions differ greatly. The validation of the crop models over different environments also improves the ability to estimate effects of changes in climate. Furthermore, because management practices, such as the choice of varieties, planting date, fertilizer application and irrigation may be varied in the models, they permit experiments that simulate adjustments by farmers and agricultural systems to climate change.

Physiological effects of CO_2. Most plants growing in experimental environments with increased levels of atmospheric CO_2 exhibit increased rates of net photosynthesis (*i.e.* total photosynthesis minus respiration) and reduced stomatal opening.[21,22] By so doing, CO_2 reduces transpiration per unit leaf area while enhancing photosynthesis. Thus, it often improves water-use efficiency (the ratio of crop biomass accumulation or yield to the amount of water used in evapotranspiration). The crop models used in this study account for the beneficial physiological effects of increased atmospheric CO_2 concentrations on crop growth and water use.[24–27]

Limitations of crop growth models. The crop growth models embody a number of simplifications. For example, weeds, diseases and insect pests are assumed to be controlled, there are no problem soil conditions (*e.g.* high salinity or acidity), and there are no extreme weather events such as heavy storms. The crop models simulate the current range of agricultural technologies available around the world. They do not include induced improvements in such technology, but may be used to test the effects of some potential improvements, such as varieties with higher thermal requirements and the installation of irrigation systems.

Yield simulations. Crop modelling simulation experiments were performed at 112 sites in 18 countries for the baseline climate (1951–80) and the GCM-doubled

[20] S. Otter-Nacke, D. C. Godwin and J. T. Ritchie, *Testing and Validating the CERES-Wheat Model in Diverse Environments*, AGGRISTARS YM-15-00407, Johnson Space Center No. 20244, Houston, 1986.

[21] B. Acock and L. H. Allen, Crop responses to elevated carbon dioxide concentrations, in B. R. Strain and J. D. Cure, (eds.), *Direct Effects of Increasing Carbon Dioxide on Vegetation*, US Department of Energy, Washington, DC, DOE/ER-0238, 1985, pp. 33–97.

[22] J. D. Cure, Carbon dioxide doubling responses: a crop survey, in B. R. Strain and J. D. Cure, (eds.), *Direct Effects of Increasing Carbon Dioxide on Vegetation*, US Department of Energy, Washington, DC, DOE/ER-0238, 1985, pp. 33–97.

[23] B. A. Kimball, Carbon dioxide and agricultural yield. An assemblage and analysis of 430 prior observations, *Agron. J.*, 1983, **75**, 779–788.

[24] H. H. Rogers, G. E. Bingham, J. D. Cure, J. M. Smith and K. A. Surano, Responses of selected plant species to elevated carbon dioxide in the field, *J. Environ. Qual.*, 1983, **12**, 569–574.

[25] J. D. Cure and B. Acock, Crop responses to carbon dioxide doubling: a literature survey, *Agric. Forest Meteorol.*, 1986, **38**, 127–145.

[26] L. H. Allen, Jr., K. J. Boote, J. W. Jones, P. H. Jones, R. R. Valle, B. Acock, H. H. Rogers and R. C. Dahlman, Response of vegetation to rising carbon dioxide: photosynthesis, biomass and seed yield of soybean, *Global Biogeochem. Cycles*, 1987, **1**, 1–14.

[27] R. M. Peart, J. W. Jones, R. B. Curry, K. Boote and L. H. Allen, Jr., Impact of climate change on crop yield in the south-eastern USA, in J. B. Smith and D. A. Tirpak (eds.), *The Potential Effects of Global Climate Change on the United States*, US Environmental Protection Agency, Washington, DC, 1989.

CO_2 climate change scenarios, with and without the physiological effects of CO_2. This involved the following tasks:

Deriving Estimates of Potential Yield Changes

Aggregation of site results. Crop model results for wheat, rice, maize and soybean from all sites and 18 countries were aggregated by weighting regional yield ranges (based on current production) to estimate change in national yields. The regional yield estimates represent the current mix of rainfed and irrigated production, the current crop varieties, nitrogen management and soils. Since the site results relate to regions that account for about 70% of the world's grain production,[28] the conclusions concerning world production total contained in this report are believed to be adequately substantiated.

The World Food Trade Model

The estimates of climate-induced changes in yields were used as inputs to a dynamic model of the world food system (the Basic Linked System) in order to assess the possible impacts on the future levels of food production, food prices and the number of people at risk from hunger. Impacts were assessed for the year 2060, with population growth, technology trends and economic growth projected to that year. Assessments were first made assuming no climate change and subsequently with the climate change scenarios described above. The difference between the two assessments is the climate-induced effect. A further set of assessments examined the efficacy of a number of adaptations at the farm level in mitigating the impact and the effect on future production of liberalizing the world food trade system, and of different rates of growth of economy and population.

The Basic Linked System (BLS) consists of linked national models. The BLS was designed at the International Institute for Applied Systems Analysis for food policy studies, but it also can be used to evaluate the effect of climate-induced changes in yield on world food supply and agricultural prices. It consists of 20 national and/or regional models that cover around 80% of the world food trade system. The remaining 20% is covered by 14 regional models for the countries that have broadly similar attributes (*e.g.* African oil-exporting countries, Latin American high income exporting countries, Asian low-income countries). The grouping is based on country characteristics such as geographical location, income per capita and the country's position with regard to net food trade (Figure 1).

The BLS is a general equilibrium model system, with representation of all economic sectors, empirically estimated parameters and no unaccounted supply sources or demand sinks.[8] In the BLS, countries are linked through trade, world market prices and financial flows. It is a recursively dynamic system: a first round of exports from all countries is calculated for an assumed set of world prices, and international market clearance is checked for each commodity. World prices are then revised using an optimizing algorithm and again transmitted to the national model. Next, these generate new domestic equilibria and adjust net exports. This

[28] FAO, *Sixth World Food Survey*, United Nations Food and Agriculture Organization, Rome, 1986.

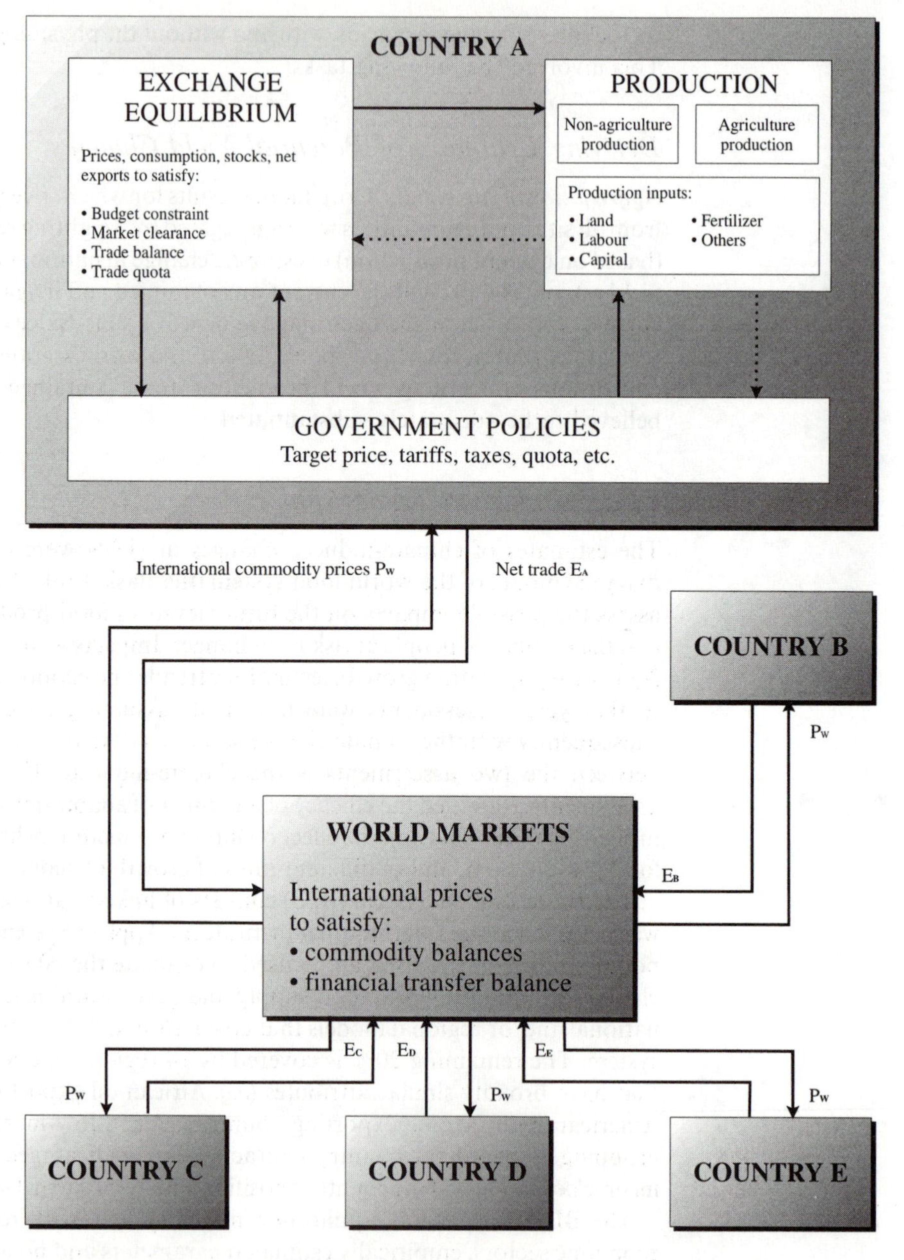

Figure 1 The Basic Linked System – relationships between country components and world markets. Arrows to countries represent international commodity prices; arrows to world markets represent net trade[6]

process is repeated until the world markets are cleared of all commodities. At each stage of the reiteration domestic markets are in equilibrium. This process yields international prices as influenced by governmental and inter-governmental agreements.

The system is solved in annual increments, simultaneously for all countries. Summary indicators of the sensitivity of the world system used in this report include world cereal production, world cereal prices and prevalence of world population at risk from hunger (defined as the population with an income insufficient to produce or procure their food requirements).

The BLS does not incorporate any climate relationships *per se*. Effects of changes in climate were introduced to the model as changes in average national or regional yield per commodity as estimated above. Ten commodities are included in the model: wheat, rice, coarse grains (*e.g.* maize, millet, sorghum and barley), bovine and ovine meat, dairy products, other animal products, protein feeds, other food, non-food agriculture and non-agriculture. In this context, however, consideration is limited to the major grain food crops.

The Set of Model Experiments

The results described in this article consider the following scenarios:

The reference scenario. This involved projection of the agricultural system to the year 2060 with no effects of climate change on yields and with no major changes in political or economic context of the world food trade. It assumed:

- UN medium population estimates (10.2 billion by 2060).[29,30]
- 50% trade liberalization in agriculture introduced gradually by 2020.
- Moderate economic growth (ranging for 3.0% per year in 1980–2000 to 1.1% per year in 2040–2060).
- Technology is projected to increase yields over time (cereal yields for world total, developing countries and developed countries are assumed to increase annually by 0.7%, 0.9% and 0.6%, respectively).
- No changes in agricultural productivity due to climate change.

Three climate change scenarios. These are projections of the world food trade system including the effects on agricultural yields under different climate scenarios (the '$2 \times CO_2$ scenarios' for the GISS, GFDL, and UKMO GCMs). The food trade simulations for these three scenarios were started in 1990 and assumed a linear change in yields until the double CO_2 concentration was reached in 2060. Simulations were made both with and without the physiological effects of 555 ppmv CO_2 on crop growth and yield for the equilibrium yield estimates. In these scenarios, internal adjustments in the model occur, such as increased agricultural investment, reallocation of agricultural resources according to economic returns and reclamation of additional arable land as an adjustment to higher cereal prices, based on shifts in comparative advantage among countries and regions.

Scenarios including the effect of farm-level adaptations. The food trade model was first run with yield changes assuming no external adaptation to climate change and was then re-run with different climate-induced changes in yield, assuming a range of farm-level adaptations. These included such measures as altering planting dates and crop varieties and the use of different amounts of irrigation and fertilizer. Two adaptation levels to cope with potential effects on

[29] United Nations, *World Population Prospects 1988*, United Nations, New York, 1989.

[30] International Bank for Reconstruction and Development/World Bank, *World Bank Population Projections*, Johns Hopkins University Press, Baltimore, 1990.

yield and agriculture were considered. Adaptation Level 1 included those adaptations at the farm level that would not involve any major changes in agricultural practices. It thus took account of changes in planting date, amounts of irrigation and the choice of crop varieties that are currently available. Adaptation Level 2 encompassed, in addition to the former, major changes in agricultural practices, such as large shifts of planting date, the availability of new cultivars, extensive expansion of irrigation and increased fertilizer application. This level of adaptation would be likely to involve policy changes both at the national and international level and significant costs. However, policy, cost and water were not studied explicitly.

Scenarios of different future trade, economic and population growth. A final set of scenarios assumed changes to the world tariff structure and different rates of growth of economy and population. As with previous experiments, these were conducted both with and without climate change impacts. These scenarios included:

- Full trade liberalization. Full trade liberalization in agriculture introduced gradually by 2020.
- Lower economic growth (ranging from 2.7% per year in 1980–2000 to 1.0% in 2040–2060). Global GDP in 2060 is 10.3% lower than the reference scenario, 11.2% lower in developing countries and 9.8% lower in developed countries.
- Low population growth. UN low population estimates (*ca.* 8.6 billion by 2060).

Effects on Yields

The results show that climate change scenarios excluding the direct physiological effects of CO_2 predict decreases in simulated yields in many cases, while the direct effects of increasing atmospheric CO_2 mitigate the negative effects primarily in mid- and high-latitudes. The differences between countries in yield responses to climate change are related to differences in current growing conditions. At low latitudes crops are grown nearer the limits of temperature tolerance and global warming may subject them to higher stress. In many mid- and high-latitude areas, increasing temperatures may benefit crops otherwise limited by cold temperatures and short growing seasons in the present climate.

The primary causes of decreases in yield are:

1. Shortening of the growing period (especially the grain filling stage) of the crop. This occurs at some sites in all countries.
2. Decreases in water availability. Depletion of soil water is increased by greater evapotranspiration and, in some cases, a decrease in precipitation in the climate change scenarios. This occurred in Argentina, Brazil, Canada, France, Japan, Mexico and USA.
3. Poor vernalization. Some temperate cereal crops require a period of low temperature in winter to initiate the flowering process. Inadequate vernalization results in low flower bud initiation and ultimately in reduced yields. This caused decreases in yields in winter wheat yields in some sites in Canada and the former USSR.

Figure 2 Change in crop yields under three equilibrium climate change scenarios with the direct effects of CO_2 taken into account[6]

GISS 2xCO2

WITH DIRECT CO2 EFFECTS

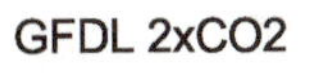

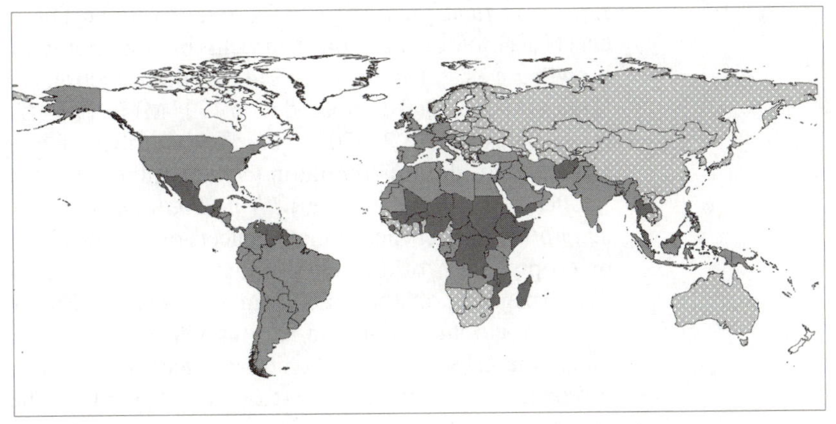

UKMO 2xCO2

WITH DIRECT CO2 EFFECTS

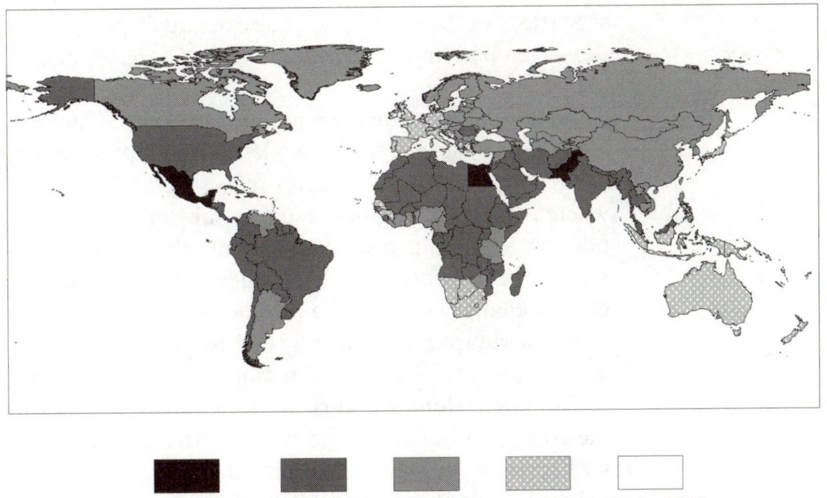

Figure 2 shows estimated potential changes in average national crop yields for the GISS, GFDL and UKMO 2 × CO_2 climate change scenarios, allowing for the direct effects of CO_2 on plant growth. Latitudinal differences are apparent in all the scenarios. Higher latitude changes are less negative or even positive in some cases, while lower latitude regions indicate more negative effects of climate change on agricultural yields.

The GISS and GFDL climate change scenarios produced yield changes ranging from +30 to −30%. The GISS scenario is, in general, more detrimental than the GFDL for crop yields in parts of Asia and South America, while GFDL results in more negative yields in the USA and Africa and less positive results in the former USSR. The UKMO climate change scenario, which has the greatest warming (5.2 °C global surface air temperature increase) suggests yield declines almost everywhere (up to −50% in Pakistan).

Effects on World Food Trade

Effects on food production. The future without climate change. Assuming no effects of climate change on crop yields but that population growth and economic growth are as stated above, world cereal production is estimated at 3286 million tonnes (Mt) in 2060 compared with 1795 Mt in 1990. Cereal prices are estimated at an index of 121 (1970 = 100). The number of people at risk of hunger is estimated at about 640 million (*cf.* 530 million estimated in 1990).

Effects of climate change with internal adjustment in the model but without adaptation. Under the estimated effects of climate change and atmospheric CO_2 on crop yields, world cereal production is estimated to decrease between 1 and 7% depending on the GCM climate scenario (Figure 3). Under the UKMO scenario, global production is estimated to decrease by more than 7%, while under the GISS scenario (which assumes lower temperature increases) cereal production is estimated to decrease by just over 1%. The largest negative changes occur in developing countries, averaging −9% to −11%. By contrast, in developed countries production is estimated to increase under all but the UKMO scenario (+11% to −3%). Thus existing disparities in crop production between the developed and developing countries are estimated to grow.

Effects of climate change on production under different levels of adaptation. The study tested the efficacy of two levels of adaptation: Level 1 implies little change to existing agricultural systems reflecting farmer response to a changing climate, whereas Level 2 implies a more substantial change to agricultural systems possibly requiring resources beyond the farmer's means. Level 2 adaptation represents an optimistic assessment of world food agriculture's response to climate change conditions as predicted by the GCMs tested in this study. In each case, the adaptations were tested as possible responses to the worst climate change scenario (usually, but not always, the UKMO scenario). Changes in economics or domestic agricultural policies were beyond the scope of this study; the costs of adaptation and future water availability under the climate change scenarios were also not considered.

Level 1 adaptation included:

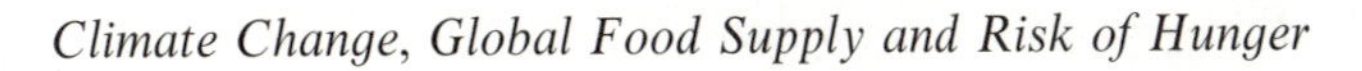

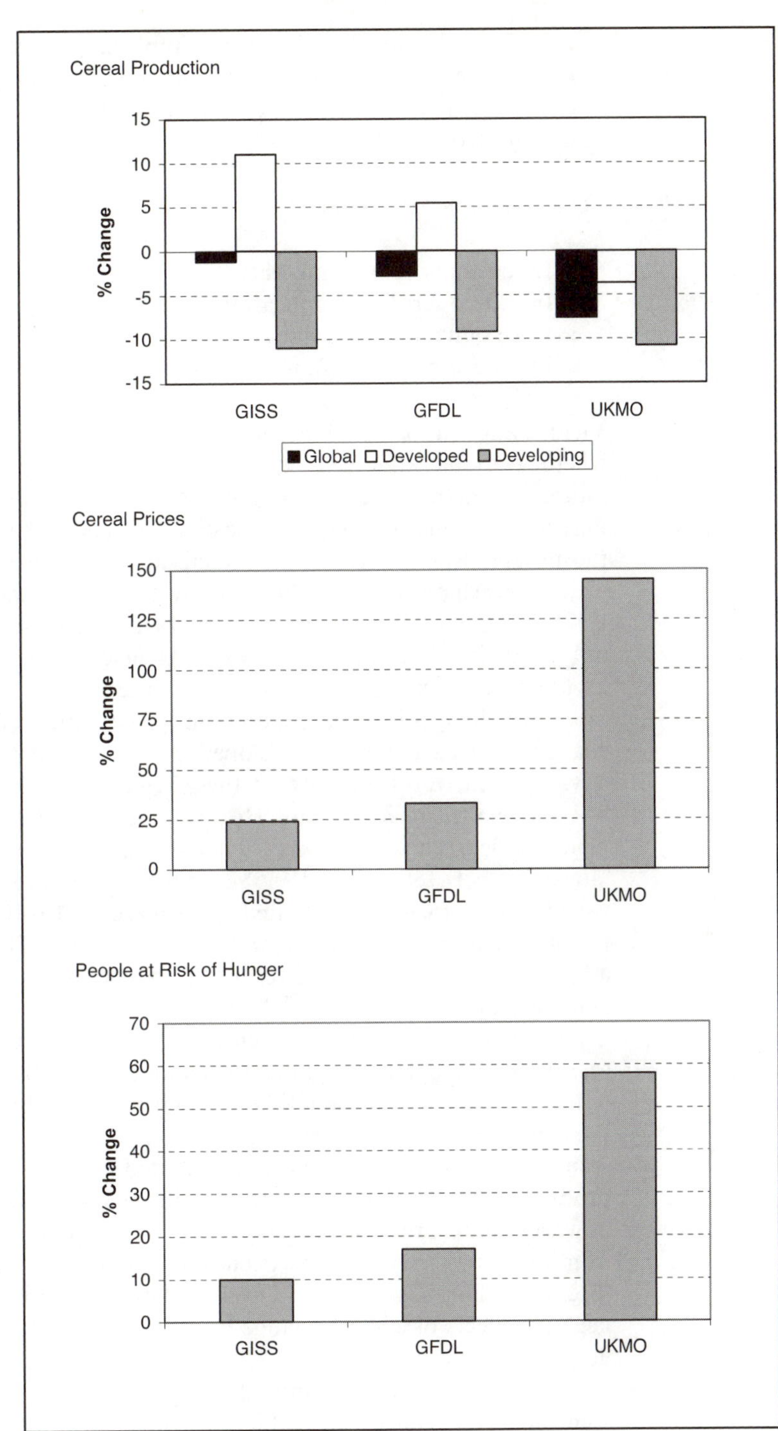

Figure 3 Change in cereal production, cereal prices and people at risk of hunger in 2060. The reference case for 2060 assumes no climate change. Cereal production = global 3286 Mt, developed 1449 Mt, developing 1836 Mt; cereal prices 1970 = 100; and 641 million people at risk of hunger[6]

- Shifts in planting date that do not imply major changes in the crop calendar;
- Additional application of irrigation water to crops already under irrigation;
- Changes in crop variety to currently available varieties better adapted to the projected climate.

Level 2 adaptation included:

- Large shifts in planting date;
- Increased fertilizer application;
- Development of new varieties;
- Installation of irrigation systems.

Yield changes for both adaptation levels were based on crop model simulations where available, and were extended to other crops and regions using the estimation methods described above. The adaptation estimates were developed only for the scenarios including the direct effects of CO_2 as these were judged to the most realistic. The two levels of adaptation estimates for the UKMO scenario were also examined. With the high level of global warming projected by the UKMO climate change scenario, neither Level 1 nor Level 2 Adaptation mitigated climate change effects on crop yields in most countries.

Adaptation Level 1. Figure 4a shows the effects of Level 1 adaptation on estimated changes in cereal production. These largely offset the negative climate change induced effects in developed countries, improving their comparative advantage in world markets. In these regions cereal production increases by 4–14% over the reference case. However, developing countries are estimated to benefit little from adaptation (−9% to −12%). Averaged global production is altered by between 0% and −5% from the reference case. As a consequence, world cereal prices are estimated to increase by 10–100% and the number of people at risk from hunger by *ca.* 5–50% (Figure 5). This indicates that Level 1 adaptations would have relatively little influence on reducing the global effects of climate change.

Adaptation Level 2. More extensive adaptation virtually eliminates negative cereal yield impacts at the global level under the GISS and GFDL scenarios and reduces impacts under the UKMO scenario by one third (Figure 4b). However, the decrease in the comparative advantage of developing countries under these scenarios leads to decreased areas planted to cereals in these areas. Cereal production in developing countries still decreases by around 5%. Globally, however, cereal prices increase by only 5–35%, and the number of people at risk from hunger is altered by between −2% and +20% from the reference case (Figure 5). This suggests that Level 2 adaptations are required to mitigate the negative effects of climate change but that these still do not eliminate them in developing countries.

Net imports of cereals into developing countries will increase under all scenarios. The change in cereal imports is largely determined by the size of the assumed yield changes, the change in relative productivity in developed and developing regions, the change in world market prices and changes in incomes of developing countries. Under the GISS climate scenario productivity is depressed

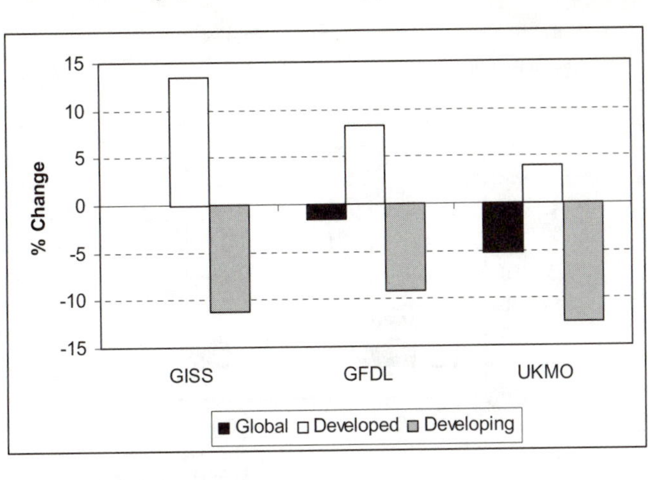

Figure 4a Change in cereal production under three equilibrium climate change scenarios in 2060 assuming implementation of Adaptation Level 1 (AD1)[6]

Figure 4b Change in cereal production under three equilibrium climate change scenarios in 2060 and assuming the implementation of Adaptation Level 2 (AD2)[6]

largely in favour of developed countries, resulting in pronounced increases of net cereal imports into developing countries. Under the UKMO scenario large cereal price increases limit the increase of exports to developing countries. Consequently, despite its beneficial impact for developed countries, the Adaptation Level 1 scenarios show only small improvements for developing countries as compared to the corresponding impacts without such adaptation.

Effects of climate change assuming full trade liberalization and lower economic and population growth rates. *Full trade liberalization.* Assuming full trade liberalization in agriculture by 2020 provides for more efficient resource use and leads to 3.2% higher value added in agriculture globally and a 5.2% higher agriculture GDP in developing countries (excluding China) by 2060 compared with the reference case. This policy change results in almost 20% fewer people at risk from hunger. Global cereal production is increased by 70 Mt, with most of the production increases occurring in developing countries. Global impacts due to climate

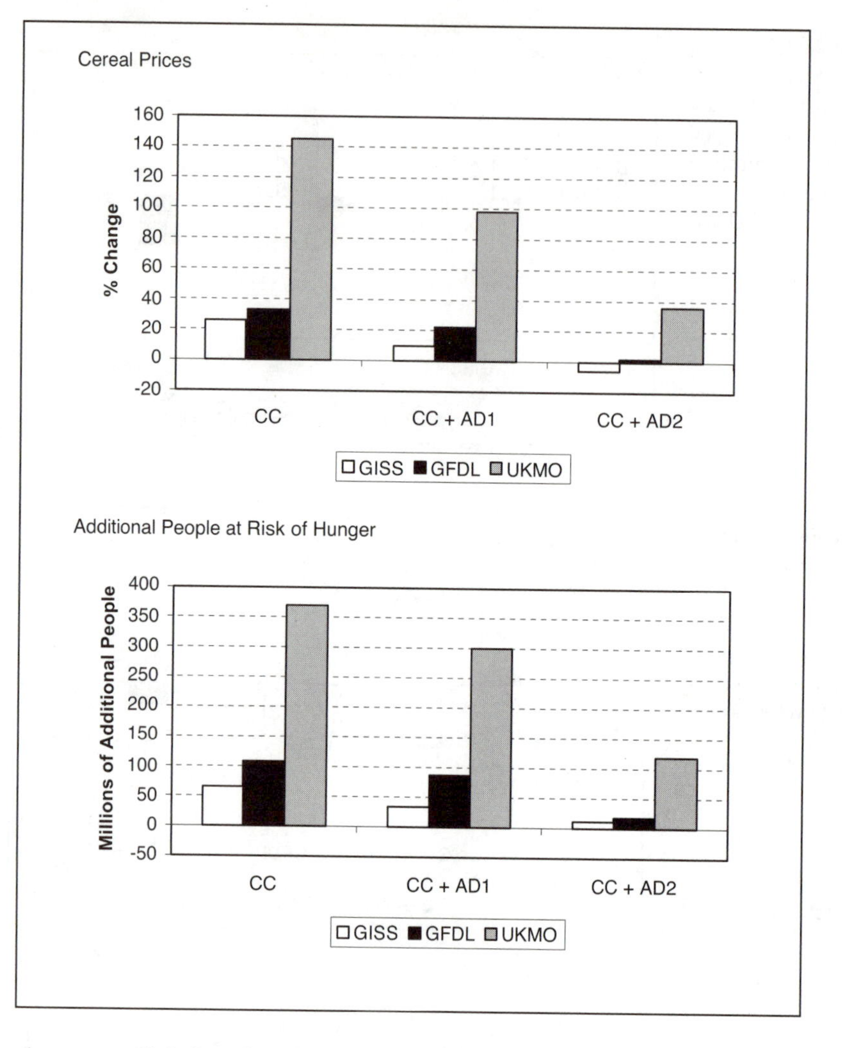

Figure 5 Change in cereal prices and people at risk of hunger for climate change scenarios (CC) and with Adaptation Levels 1 and 2 (AD1 and AD2)[6]

change are slightly reduced under most climatic scenarios, with enhanced gains in production occurring in developed countries but losses in production being greater in developing countries. Price increases are reduced slightly from what would occur without full trade liberalization and the number of people at risk from hunger is reduced by about 100 million.

Reduced rate of economic growth. Estimates were also made of impacts under a lower economic growth scenario (10% lower than reference). Lower economic growth results in a tighter supply situation, higher prices and more people below the hunger threshold. Prices are 10% higher and the number of people at risk from hunger is 20% greater. The effect of climate change on these trends is generally to reduce production, increase prices and increase the number of people at risk from hunger by the same ratio as is the case with a higher economic growth rate, but the absolute amounts of change are greater.

Altered rates of population growth. The largest impact of any of the policies considered would result from an accelerated reduction in population growth in developing countries. Simulations based on rates of population growth according to UN Low Estimates result in a world population about 17% lower in year 2060 as compared with the UN Mid Estimates used in the reference run. The corresponding reduction in the developing countries (excluding China) would be about 19.5% from 7.3 to 5.9 billion. The combination of higher GDP/capita (about 10%) and lower world population produces an estimated 40% fewer people at risk from hunger in the year 2060 compared with the reference scenario.

Even under the most adverse of the three climate scenarios (UKMO) the estimated number of people at risk from hunger is some 10% lower that that estimated for the reference case without any climate change. Increases in world prices of agricultural products, in particular cereals, under the climate change scenarios employing the low population growth projection are around 75% of those using the UN medium estimate.

3 Recently Estimated Effects for Climate Scenarios from Higher Resolution GCMs with New Socio-economic Projections

Since the mid-1990s the spatial resolution of GCMs has increased and their simulation of air–ocean interactions and other feedback mechanisms has improved. This has substantially enhanced the accuracy of their projections of climate change resulting from greenhouse gas forcing. Many are also transient in nature and are capable of producing time-dependent scenarios, thus enabling the evaluation of climate change impacts at several different time horizons throughout this century.

As the climate models have evolved, becoming ever more complex, many of the assumptions and models used in the assessment of agro-climatic impacts have also been revised. This has happened for several reasons; for example, in response to the developments made by the climate modelling community, or due to advances in understanding in both the natural and social sciences; and to dramatic changes in the economic and political structure of many of the world's major food producing regions. While it is still generally accepted that, for the foreseeable future, cereal yields will increase globally every year by nearly 1% due to technological advances, the latest estimates of the potential benefit of the CO_2 fertilization effect are much lower than was first thought in the early nineties.[32,33] Countering this reduction in adaptive capacity the latest population projections suggest a slowing down in the growth rate of the global population. Early studies projected that there would be more than 10 billion people in the world by 2060.[29] It is now thought that the 10 billion people mark will not be passed until the

[31] M. L. Parry, M. T. J. Livermore, C. Rosenzweig, A. Iglesias and G. Fischer, Stabilisation – a way of securing future world food supply?, in *Stabilisation of Atmospheric Carbon Dioxide: Its Effects on Climate Change Impacts*, Department of the Environment, Transport and the Regions/UK Meteorological Office, 1999, pp. 18–19.

[32] J. Reilly, W. Baethgen, F. E. Chege, S. C. van de Geijn, Lin Erda, A. Iglesias, G. Kenny, D. Patterson, J. Rogasik, R. Rötter, C. Rosenzweig, W. Sombroek and J. Westbrook, Agriculture in a changing climate: impacts and adaptation, in *Changing Climate: Impacts and Response Strategies*, Report of Working Group II of the Intergovernmental Panel on Climate Change, 1996.

[33] R. Darwin and D. Kennedy, Economic effects of CO_2 fertilization of crops: transforming changes in yield into changes in supply, *Environ. Modeling Assess.*, 2000, **5**, 157–168.

middle of the 2080s.[34]

The end result is that the estimates of future warming have been greatly reduced from more than 5 °C by 2060 to less than 3.5 °C by the middle of the 2080s. In response to this and changes in the environmental and socio-economic systems the latest estimates regarding future world food security are far more modest. Changes in yield now range from −10% to +10%. While much smaller than the earlier estimates[6,35] they are still considerable impacts when seen in the context of the world food trade market.

The research method has also changed slightly. In the latest suite of experiments the crop models were run for current climate conditions and for three future climate conditions predicted by the Hadley Centre's GCMs known as HadCM2.[36,37] All climate change scenarios are based on an IS92a-type forcing (one which assumes greenhouse gas emissions stem from a 'business-as-usual' future in economic and social terms).

Estimated Effects on Yields

Figures 6a–c show the estimated potential changes in average national grain crop yields for the four HadCM2 and one HadCM3 climate change scenarios, allowing for the direct effects of CO_2 on plant growth. The maps are created from the nationally averaged yield changes for wheat, rice and maize. Regional variations within countries are not shown.

The latitudinal variations in crop yields illustrated in Figures 6a–c are mainly due to differences in current growing conditions. Under the HadCM2 scenario, in many mid- and high-latitude areas, where current temperature regimes are low, the increase in surface temperatures tends to lengthen the growing season thus increasing yields. This potentially beneficial effect is not evident under the HadCM3 scenario. The intensified polar warming experienced under HadCM3 is so great that the threshold concerning positive effects of warmer temperatures at higher latitudes is exceeded and a decrease in yields occurs in some of these regions.

Another difference evident from Figures 6a–c is that, while the area most adversely affected under HadCM2 is the Indian subcontinent, under HadCM3 it is western Africa and the USA.

Estimated Effects on Food Production, Food Prices and Risk of Hunger

The reference scenario (the future without climate change). Assuming no effects of climate change on crop yields and current trends in economic and population

[34] E. Bos, T. My, E. M. Vu and R. A. Bulatao, *World Population Projections 1994–95: Estimates and Projections with Related Demographic Statistics*, World Bank, Johns Hopkins University Press, New York, 1994.

[35] C. Rosenzweig, M. L. Parry, G. Fischer and K. Frohberg, *Climate Change and World Food Supply*, Research Report No. 3, Environmental Change Unit, University of Oxford, Oxford, 1993.

[36] J. F. B. Mitchell, T. C. Johns, J. M. Gregory and S. Tett, Climate response to increasing levels of greenhouse gases and sulphate aerosols, *Nature*, 1995, **376**, 501–504.

[37] M. Hulme, J. Mitchell, W. Ingram, T. Johns, M. New and D. Viner, Climate Change Scenarios for Global Impacts Studies, *Global Environ. Change*, 1999, **9**, (4), s3–s19.

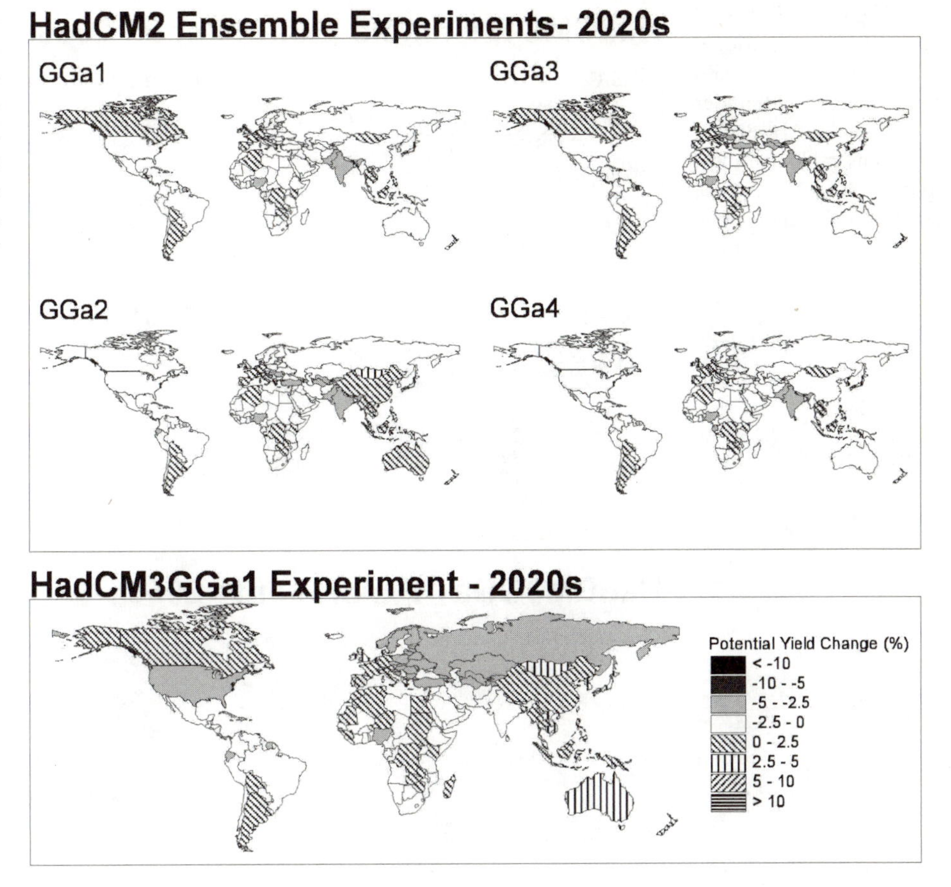

Figure 6a Potential changes (%) in national cereal yields for the 2020s (compared with 1990) under the four HadCM2 ensemble members (GGa1-4) and the single HadCM3 climate change scenarios[4]

growth rates, world cereal production is estimated at 4012 million tonnes (Mt) in the 2080s (~1800 Mt in 1990).

Cereal prices are estimated at an index of 92.5 (1990 = 100) for the 2080s, thus continuing the trend of falling real cereal prices over the last 100 years. This occurs because the BLS standard reference scenario has two phases of price development. Between 1990 and 2020, while trade barriers and protection are still in place but are being reduced, there are increases in relative prices due to the increases in demand brought about by the growing world population. However, after 2020, by which time a 50% liberalization of trade has been realized, prices begin to fall again. This has obvious ramifications for the number of hungry people which is now estimated at about 300 million or about 3% of total population in the 2080s (~521 million in 1990, about 10% of total current population).

Effects of climate change. Global effects. Changes in cereal production, cereal prices, and people at risk of hunger estimated for the HadCM2 climate change scenarios (with the direct CO_2 effects taken into account) show that world is

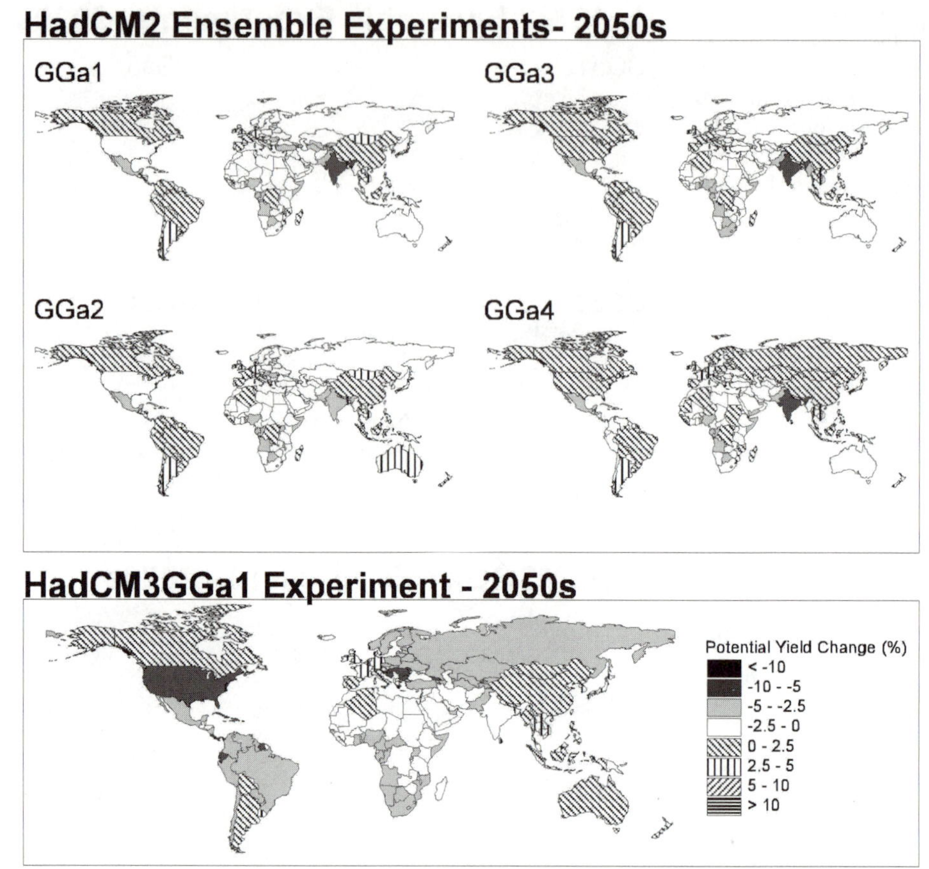

Figure 6b Potential changes (%) in national cereal yields for the 2050s (compared to 1990) under the four HadCM2 ensemble members (GGa1-4) and the single HadCM3 climate change scenarios[4]

generally able to feed itself in the next millennium. Only a small detrimental effect is observed on cereal production, manifested as a shortfall on the reference production level of around 100 Mt (−2.1%) by the 2080s (±10 Mt depending on which HadCM2 climate simulation is selected). In comparison, HadCM3 produces a greater disparity between the reference and climate change scenario – a reduction of more than 160 Mt (about −4%) by the 2080s (Figure 7a).

Reduced production leads to increases in prices. Under the HadCM2 scenarios cereal prices increase by as much as 17% (±4.5%) by the 2080s (Figure 7b). The greater negative impacts on yields projected under HadCM3 are carried through the economic system with prices estimated to increase by about 45% by the 2080s. In turn these production and price changes are likely to affect the number of people with insufficient resources to purchase adequate amounts of food. Estimations based upon dynamic simulations by the BLS show that the number of people at risk of hunger increases, resulting in an estimated additional 90 million people in this condition due to climate change (above the reference case of ~250 million) by the 2080s (Figure 7c). The HadCM3 results are again more extreme, falling outside the HadCM2 range with an estimated 125+ million

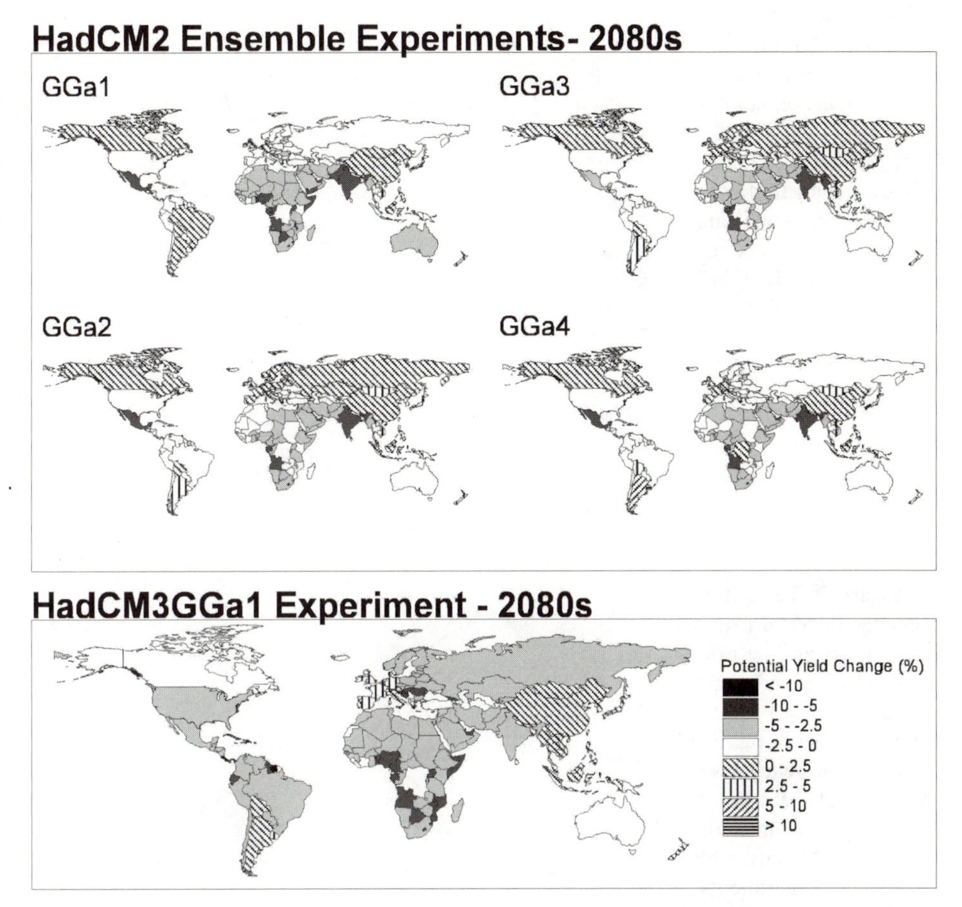

Figure 6c Potential changes (%) in national cereal yields for the 2080s (compared to 1990) under the four HadCM2 ensemble members (GGa1-4) and the single HadCM3 climate change scenarios[4]

additional people at risk of hunger by the 2080s. All BLS experiments allow the world food system to respond to climate-induced supply shortfalls of cereals and higher commodity prices through increases in production factors (cultivated land, labour, and capital) and inputs such as fertilizer.

Regional effects. The global estimates presented above mask important regional differences in impacts. For example, under the HadCM2 scenarios yield increases at high and high mid-latitudes lead to production increases in these regions, a trend that may be enhanced due to the greater adaptive capacity of countries here. Both Canada and Europe are good examples of this. In contrast, yield decreases at lower latitudes, and in particular in the arid and sub-humid tropics, lead to production decreases and increases in the risk of hunger, effects that may be exacerbated where adaptive capacity is lower than the global average.

Under the HadCM2 scenarios the largest negative changes occur in developing regions, which on average varies between −3.5% and −16.5%, though the extent of decreased production varies greatly by country depending on the projected climate. Disparities in crop production between developed and developing countries are estimated to increase. However, our results based on the

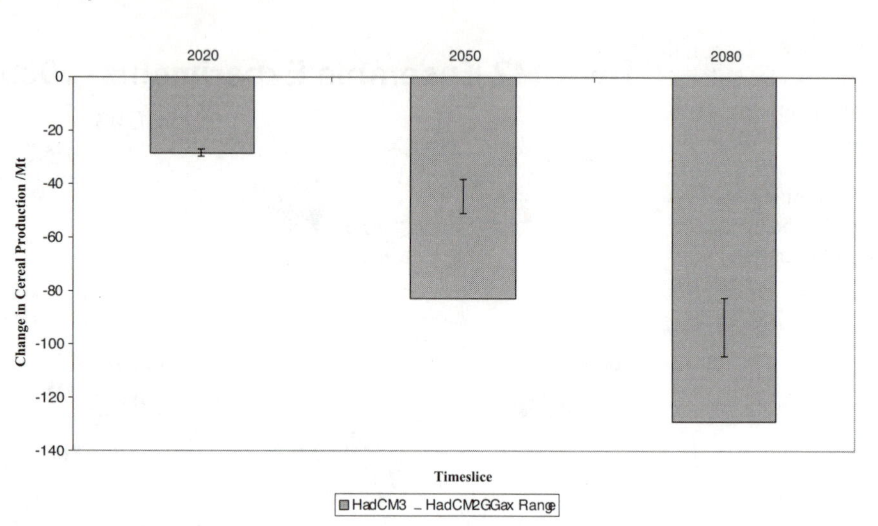

Figure 7a Changes in global cereal production (Mt). Grey blocks are the production change projected under the HadCM3 climate change scenario (compared with the reference case). Bars depict the range of change under the four HadCM2 ensemble simulations[4]

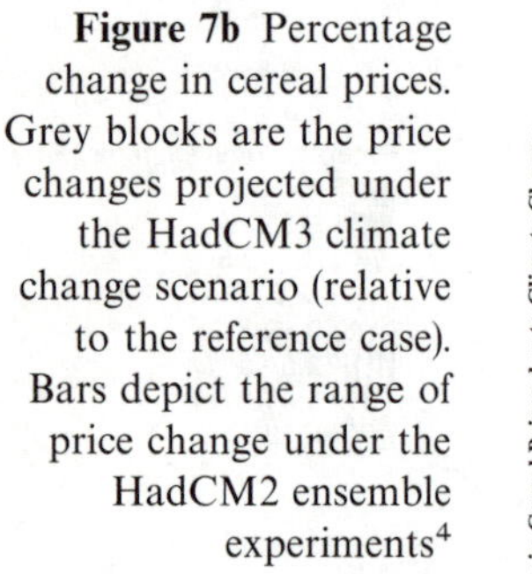

Figure 7b Percentage change in cereal prices. Grey blocks are the price changes projected under the HadCM3 climate change scenario (relative to the reference case). Bars depict the range of price change under the HadCM2 ensemble experiments[4]

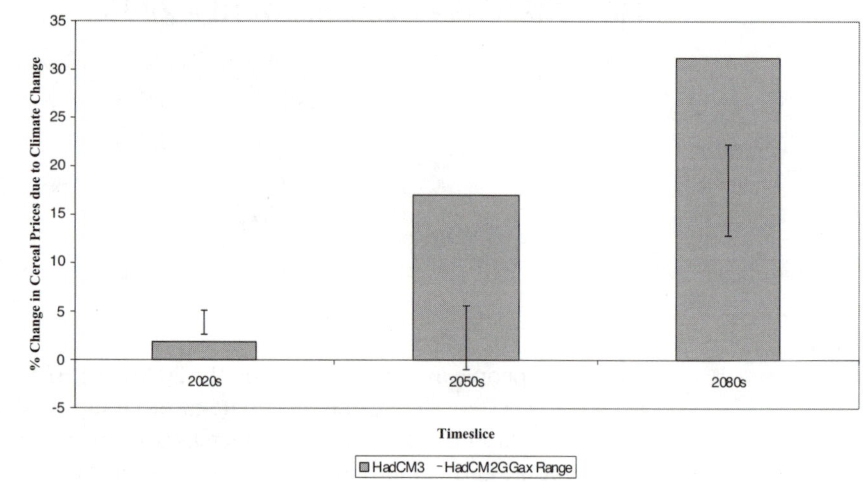

HadCM3 experiment show that the relationship between global warming and increased yields in the higher latitudes is finely balanced. Under HadCM3 the higher latitudes get warmer and drier than under the HadCM2 scenarios. The result is that for the first time negative impacts on cereal yields and production figures are evident in North America, Eastern Europe and the Russian Federation as early as the 2020s (Figure 8).

The additional range of values provided by the HadCM2 ensemble simulations suggests that developing regions may not only have to meet the challenge of a warmer world but also a more variable one. Developing regions appear less able to deal with the range of multi-decadal climate variability that is presented under

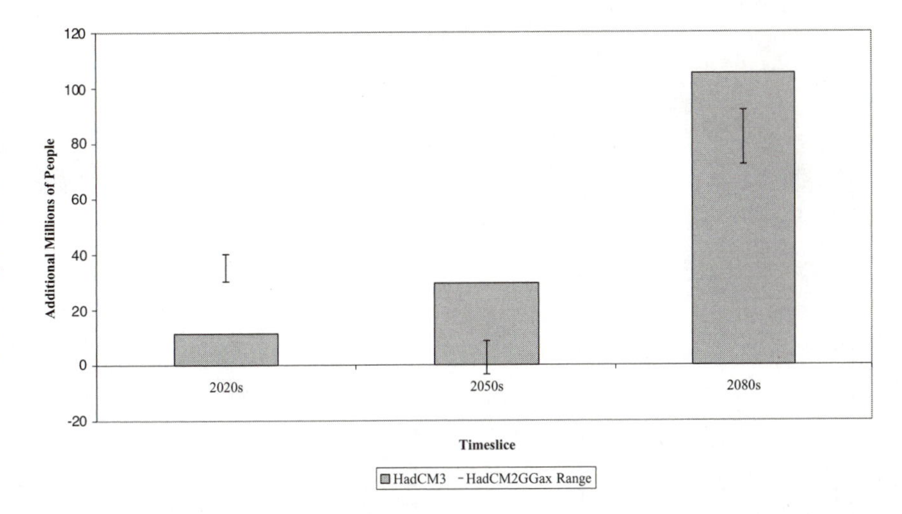

Figure 7c Global estimates of the additional number of people at risk of hunger due to climate change compared with the reference case. HadCM3 estimates are represented by the grey blocks. Bars represent the range of results under the four HadCM2 ensemble simulations[4]

the four HadCM2 scenarios. In Africa cereal productivity under the HadCM2 scenarios is estimated to be reduced by about 12% or 30 Mt ($\pm$2% depending on which HadCM2 ensemble member is chosen) from the reference case by 2080 (Figure 8). The figure for South East Asia is *ca.* 23% ($\pm$1%). As a consequence the number of people at risk of hunger in developing regions is estimated to increase: in Africa by more than one-third, while in Latin America we might expect to see a doubling over reference case levels (Figure 9).

4 Reducing Impacts by Stabilizing CO_2 Concentrations at Lower Levels

In this section we explore the implications for a range of global-scale impacts of climate change of the stabilization of CO_2 concentrations at defined levels. These stabilization scenarios (at 550 ppmv by 2150 and 750 ppmv by 2250) are among the set defined by the Intergovernmental Panel on Climate Change.[38]

Scenarios

Two stabilization scenarios (stabilizing at CO_2 concentrations of 550 ppmv and 750 ppmv) are considered, and compared with the IS92a unmitigated emissions scenario.[39] Figure 10 shows the CO_2 emissions profiles and atmospheric CO_2 concentrations consistent with the forcings used in the climate experiments (note that the Figure does not show the concentration and emissions of other greenhouse gases). There is little difference in concentrations between the two scenarios to the 2020s, but thereafter they begin to diverge. The S750 scenario

[38] IPCC, *Climate Change 2001: The Scientific Basis. Technical Summary of the Working Group I Report*, Geneva, 2001.

[39] J. Leggett, W. J. Pepper and R. J. Swart, Emissions scenarios for the IPCC: an update, in J. T. Houghton, B. A. Callander and S. K. Varney, (eds.), *Climate Change 1992: the Supplementary Report to the IPCC Scientific Assessment*, Cambridge University Press, Cambridge, 1992, 75–95.

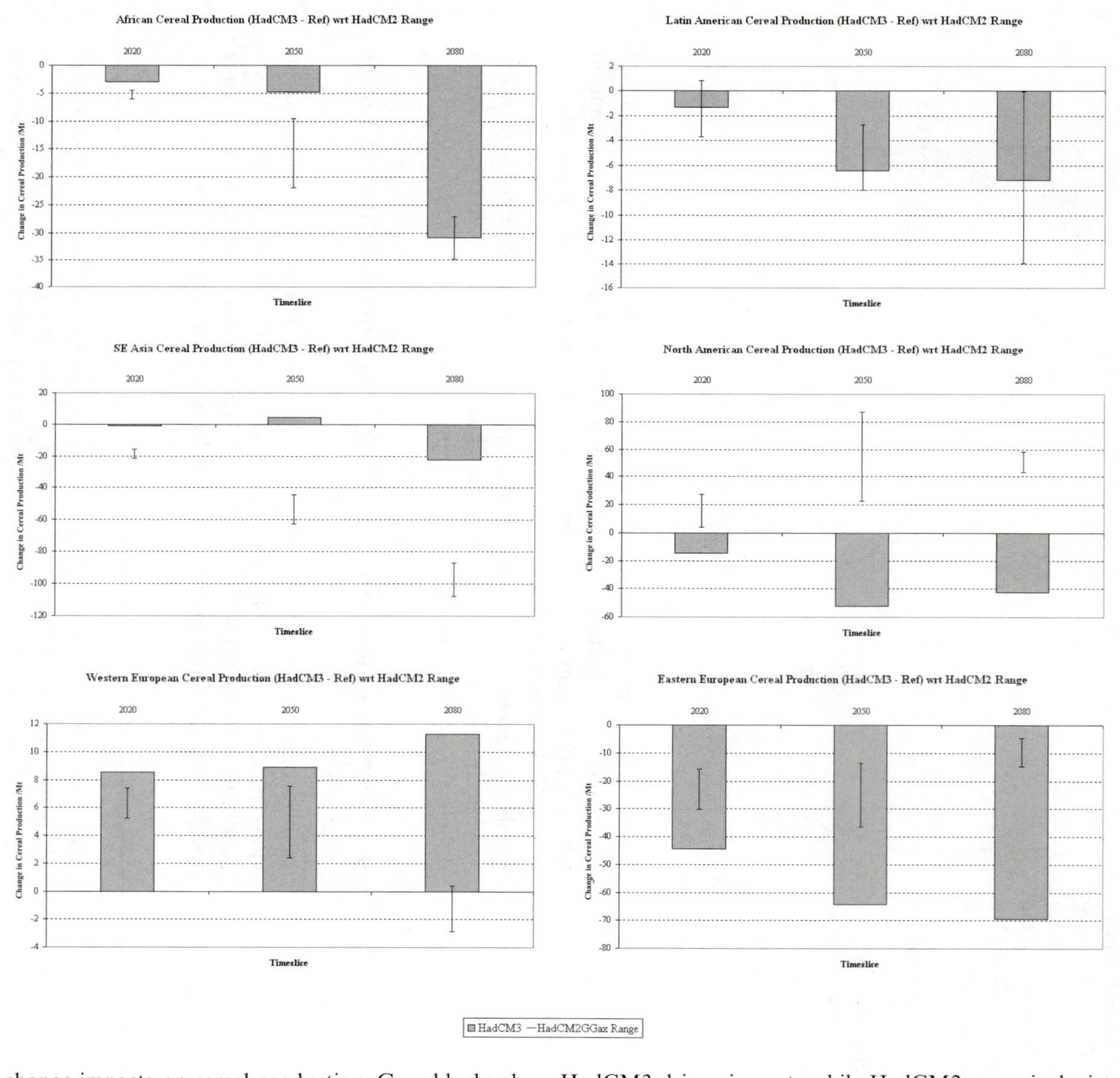

Figure 8 Regional climate change impacts on cereal production. Grey blocks show HadCM3 driven impacts while HadCM2 range is depicted by the bars[4]

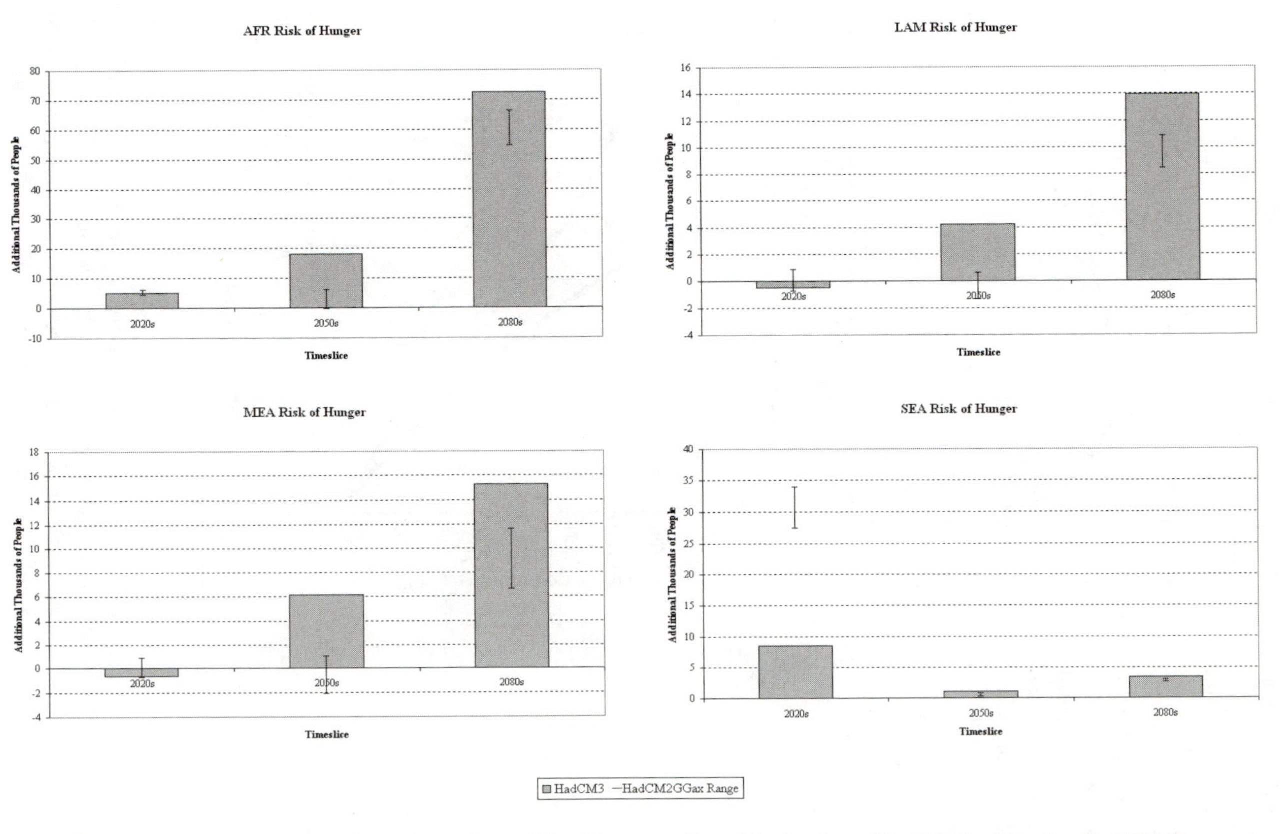

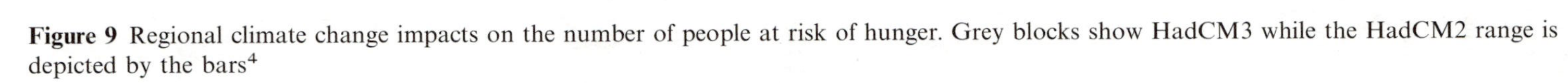

Figure 9 Regional climate change impacts on the number of people at risk of hunger. Grey blocks show HadCM3 while the HadCM2 range is depicted by the bars[4]

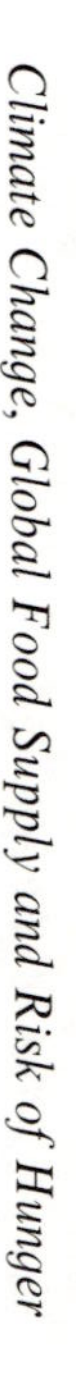

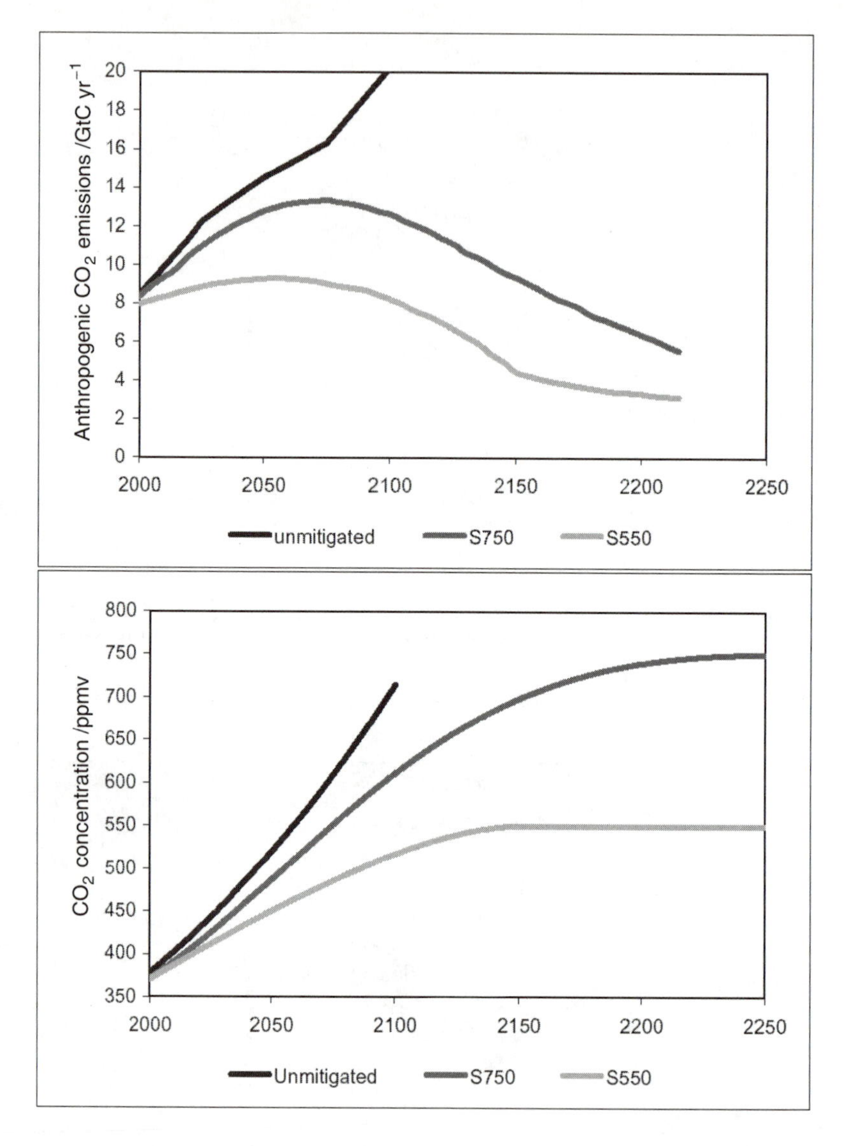

Figure 10 CO_2 emissions profiles and concentrations under the three scenarios considered: unmitigated (IS92a: top line), S750 (middle line) and S550 (bottom line). The emissions are as calculated using the Bern carbon cycle model, and are taken from IPCC[40]

stabilizes CO_2 concentrations by 2250, whilst the S550 scenario assumes stabilization occurs by 2150. Achieving stabilization at 750 ppmv and 550 ppmv, under the pathways assumed here, requires cuts in annual CO_2 emissions of around 13% and 30% respectively by 2025, relative to the 2025 emissions assumed under IS92a. We interpret these stabilization scenarios as representing actual CO_2 concentrations for the purposes of crop and vegetation modelling (*e.g.* actual CO_2 concentration reaches 750 ppmv by 2250), because there are no accepted stabilization scenarios for the other radiatively significant trace gases.

[40] IPCC, *Stabilisation of Atmospheric Greenhouse Gases: Physical, Biological and Socio-economic Implications, Intergovernmental Panel on Climate Change Technical Paper III*, Intergovernmental Panel on Climate Change, 1997.

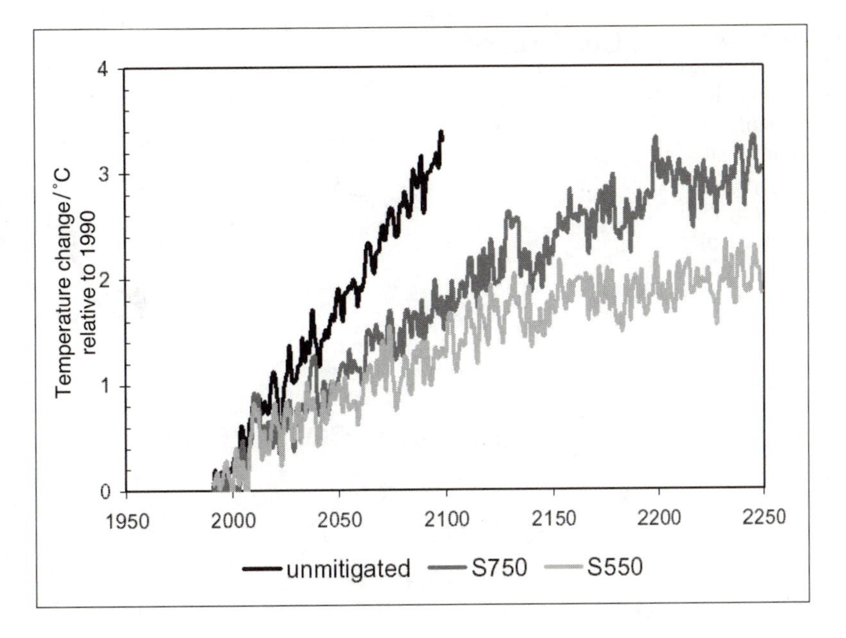

Figure 11 Global temperature change under unmitigated emissions (top line), S750 (middle line) and S550 (bottom line),[41] relative to 1990 levels

We therefore assume that all other greenhouse gas concentrations remain constant at 1990 values.

None of the scenarios included climate forcing due to changing sulfate aerosol concentrations, because the IS92a sulfate aerosol scenarios are now regarded as being too extreme and sulfate aerosol emissions under stabilization have not been estimated.

Figure 11 shows global average temperature, relative to the mean temperature under the control run with pre-industrial greenhouse gas concentrations, under the three emissions scenarios. The two stabilization scenarios diverge from the unmitigated emissions scenario at around the 2020s, but are similar to each other until at least the 2070s. By 2250, temperatures are simulated to reach about 3.3 °C and 2.3 °C above the 1961–1990 average under S750 and S550 respectively: the rise in temperature appears to stabilize around 2170 under S550, and has perhaps not quite stabilised by 2230 under S750. In other words, temperature stabilization lags behind CO_2 concentration stabilization by at least 20 years. A global temperature rise of around 2 °C, which would occur under unmitigated emissions by the 2050s, would be delayed by about 50 years under S750 and around 100 years under S550.

Effects on Yield Potential

Figure 12 shows the estimated changes in national potential grain yield by the 2080s, assuming no changes in crop cultivars, under the three emissions

[41] J. F. B. Mitchell, T. C. Johns, W. J. Ingram and J. A. Lowe, The effect of stabilising atmospheric carbon dioxide concentrations on global and regional climate change, *Geophys. Res. Lett.*, 2000, **27**, 2997–3100.

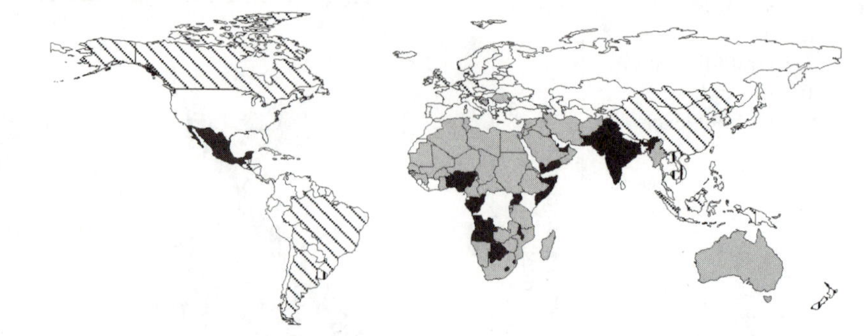

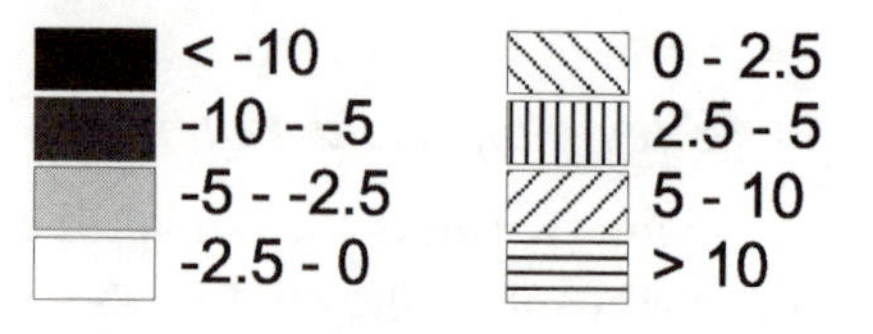

Figure 12 Changes in national cereal crop yields by the 2080s under three different emissions scenarios – unmitigated (IS92a: top map), S750 (middle map) and S550 (bottom map)[42]

Table 1 Average annual cereal production (million tonnes)[42]

	No climate change	*Unmitigated*	*S750*	*S550*
1990	1800			
2020s	2700	2670–2674	2672	2676
2050s	3500	3475	3973	3477
2080s	4000	3927	3987	3949

Notes: The estimates assume no change in crop cultivar, and come from the Basic Linked System.
The range in estimates for the unmitigated scenario represents the range between the four ensemble partners.

Table 2 Number of people at risk of hunger (millions)[42]

	No climate change	*Unmitigated*	*S750*	*S550*
1990	521			
2020s	496	521–531	546	540
2050s	312	309–321	319	317
2080s	300	369–391	317	343

Notes: The range in estimates for the unmitigated scenario represents the range between the four ensemble partners.

scenarios. Under unmitigated emissions, positive changes in mid and high latitudes are overshadowed by reductions in yield in the lower latitudes. These reductions are particularly substantial in Africa and the Indian subcontinent. However, many of the mapped changes in yield are small and indistinguishable from the effects of natural climate variability.

Stabilization at 550 ppmv produces far fewer reductions in yield, although there would still be reductions in the Indian subcontinent, most of the Pacific Islands, central America and the majority of African nations. Stabilization at 750 ppmv to a large extent produces intermediate changes. However, there are some interesting anomalies. Significant increases in yields are seen in the mid-latitudes of both hemispheres under S750 which are not replicated under S550. To a certain extent, this reflects differences in simulated regional climate – particularly precipitation – between scenarios due to natural climatic variability, but there is also a complex balance between the effects of higher temperatures, higher atmospheric CO_2 concentrations, altered rainfall and optimal growing conditions. The intermediate combination of increases in temperatures, available moisture and ambient CO_2 concentrations experienced under S750 lead in some regions to an enhancement of crop productivity that is not witnessed in the unmitigated world (which has higher CO_2 concentrations, but is warmer and with more extreme changes in moisture) or the S550 world (which does not see as large changes in temperature, moisture availability or the beneficial effects of atmospheric CO_2).

[42] N. W. Arnell, M. G. R. Cannell, M. Hulme, R. S. Kovats, J. F. B. Mitchell, R. J. Nicholls, M. L. Parry, M. T. J. Livermore and A. White, The consequences of CO_2 stabilisation for the impacts of climate change, *Clim. Change*, in press.

It should also be noted that the larger regional increases and decreases in crop yields witnessed under S550 and S750 by the 2080s fall outside the range of previously reported results from the ensemble of unmitigated HadCM2-driven experiments.[4] Table 1 summarizes global cereal production (under realistic assumptions about trade liberalization) in the absence of climate change and under the three emissions scenarios.

Implications for Food Security and Hunger

The changes in total global cereal production shown in Table 1 appear small, but can have significant effects on global food prices and the consequent risk of hunger. Food prices are simulated in the Basic Linked System, and are projected to rise relative to the baseline case with no climate change because of the lower production. This increase in prices exacerbates the stress of regional shortfalls in production, leading to an increase in the risk of hunger. More cases emerge where populations are not only unable to grow enough food due to a sustained deterioration in their resource base, but are also unable to reduce the food deficit by purchasing additional foodstuffs on the world markets because of regional inequalities in economic growth.

Table 2 shows the number of people at risk from hunger in the absence of climate change and under the three emissions scenarios. With no climate change, the number of people at risk from hunger, following historical trends, decreases from more than 500 million in 1990 to about 270 in the 2080s. This is the result of increased agricultural production due to technological advances combined with the assumption that living standards will rise while the incidence of poverty in developing countries will continue to fall. Under the unmitigated emissions scenario, it is estimated that the additional number of people at risk from hunger due to climate change would be around 20 million by the 2050s, increasing to around 80 million by the 2080s.[4] The numbers of people affected are smaller by the 2050s, largely because the effects of climate change on prices are lower at this time,[4] which is itself because at this time horizon – unlike the others – grain production increases in the United States under two of the four ensemble members. Stabilization at 750 ppmv reduces the unmitigated impacts by around 75%, while stabilization at 550 ppmv achieves a more modest mitigating reduction of around 50% in the number of additional people at risk of hunger due to climate change.

Global figures, however, hide considerable regional variations. The vast majority (~65%) of the people at additional risk of hunger in the future are in Africa. This partly reflects the greater-than-average reduction in yields, but is also due to higher levels of vulnerability caused to some extent by the lower incomes in Africa. Increasing this regional disparity, it appears that the beneficial effects of stabilization are also less in this region. Under an S750 world the additional number of people at risk of hunger is only reduced by ~30% while under an S550 future the reduction in the climate induced impact is only 20%.

5 Conclusions

Broadly, climate change seems likely to lead to increases in yield potential at mid- and high mid-latitudes, and to decreases in the Tropics and Subtropics. But there are many exceptions, particularly where increases in monsoon intensity or where more northward penetration of monsoons leads to increases in available moisture.

Risk of hunger increases generally as a result of climate change, particularly in southern Asia and Africa. However, this geographic distribution in some areas is more the result of projected increase in number of poor people in these regions (*i.e.* the exposed population) than of the regional pattern of climate change.

Much, of course, is uncertain. In particular, we are unclear about the potentially beneficial effects of elevated CO_2 on crop growth. Current estimates are based upon field experiments that have assumed near-optimal applications of fertilizer, pesticide and water, and it is possible that the actual 'fertilizing' effect of higher levels of CO_2 is less than we expect. Moreover, we have not taken into account effects of altered climate on pests and weeds, which are likely to vary greatly from one environment to another.

Although we have considered two levels of adaptation, these barely begin to capture the range of options that is open to farmers. What is, however, initially evident (and intuitively makes sense) is that the potential for adaptation is greater in more developed economies and that this, together with the generally more favourable effects of climate change on yield potential in higher rather than lower latitude regions, is likely on balance to bring more positive effects to the North and more negative effects to the South; in other words, to aggravate inequalities in development potential.

Global Environmental Changes and Human Health

MICHAEL J. AHERN AND ANTHONY J. McMICHAEL

1 Introduction

Earth's biophysical environment is a large and complex system, driven by solar energy, and is the source of nature's capital of 'goods and services'.[1] Human societies, like all other species, are not only dependent upon, but are inextricably linked with this complex system. This environment provides the basics for human survival – air, food and water. It also provides other life-supporting environmental 'goods' – for example clothing, shelter and energy – and 'services' – maintenance of the hydrological cycle and uptake of carbon dioxide and production of oxygen via plant photosynthesis.

Over many millennia, human societies have influenced the 'carrying capacity' of their local environment by occupying greater tracts of territory, modifying local environments and exploiting the local resources. These exploits varied in scale and intensity, and required a trade-off between increasing human population density and the potential weakening of the life-supporting capacity of local environments. In the past two centuries the scale of human impact on the environment has increased rapidly as human numbers have expanded and as the material- and energy-intensity of economic activity has increased. Global economic activity, for example, increased approximately twenty-fold in the last century, and whilst we remain uncertain of Earth's human 'carrying capacity'[2] demographers expect that aggregate world population, currently at 6.1 billion, will probably reach around 9 billion by 2050. Depending on subsequent scenarios, world population should stabilize by the end of the 21st century somewhere within the range of 7–15 billion.[3,4]

Concerns over environmental risks to human health were, until recently, chiefly confined to local issues, such as microbial and chemical water pollution and urban air pollution. However, regional environmental problems, including

[1] R. Costanza, R. d'Arge, R. de Groot, S. Farber, M. Grasso, B. Hannon, K. Limburg, S. Naeem, R. V. O'Neill, J. Paruelo, R. G. Raskin, P. Sutton and M. van den Belt, *Nature*, 1997, **387**, 253.

[2] J. E. Cohen, *How Many People Can the Earth Support?*, Norton, 1995.

[3] J. C. Caldwell, *Br. Med. J.*, 1999, **319**, 985.

[4] United Nations Population Division Department of Economic and Social Affairs, *World Population Prospects: the 2000 Revision*, United Nations, 2001.

Issues in Environmental Science and Technology, No. 17
Global Environmental Change

acid rain and the diffusion of persistent chlorinated organic chemicals, were recognized several decades ago. In the last two decades several global-scale environmental changes, historically unprecedented, have emerged – and have been recognized as posing substantial risks to human health. These include changing the gaseous composition of the lower and middle atmospheres; reducing productive soils on all continents; depleting ocean fisheries; over-exploiting and contaminating many of the great aquifers upon which irrigated agriculture depends; and resulting in an unprecedented rate of loss of whole species and many local indigenous populations. Anthropogenic activity is disrupting at a global level many of the biosphere's life-support systems, which provide environmental stabilization, replenishment, organic production, the cleansing of water and air, and the recycling of nutrient elements.

Assessments of these human-induced changes have concluded that humanity is now incurring a significant and increasing 'ecological deficit', evidenced by an increasing decline in global environmental and ecological resource stocks.[5,6] Today, we have reached a position where the aggregate human impact on Earth's natural capital has reached unprecedented levels, at national, regional and global scales. Some of these impacts are likely to have longer-term, and commensurately more serious, consequences for the health of human populations.

This article reviews how human activities are adversely affecting, at various physical scales, the different media of the biophysical environment, and in turn how these changes may affect human health – the focal topic of Section 2 of the article. As human population density continues to rise, with a commensurate rise in the material- and energy-intensity of economic activity, this will place pressures on many natural resources, which have finite stocks. Such pressures may lead to conflict and insecurity, with adverse consequences for human health; this important subject is reviewed in Section 3.

Often our environmental heritage and dependency have been neglected, through either a lack of motivation or simple ignorance. In today's world where human activities are having an impact at a global scale there is a need not only for society to understand this heritage, but to ensure that our dependence on the environment is managed in a sustainable manner, for both current and future generations. There is therefore a need to provide a framework for a better understanding of this heritage and dependency. In Section 4 the concept of the environment as a global public good for health is introduced, and we consider how this framework might be used to enhance the interface between research and policy, with a view to a sustainable future for all.

2 Local, Regional and Global Environmental Changes

The complexity of the biophysical environment can be seen in terms of media – air, water, soil and climate – and physical scale – local, regional, global, and cross-scale. Cross-scale recognizes that the scale of an environmental health impact may greatly exceed the scale of the initial exposure; an example is where

[5] C. D. Butler, *Global Change and Human Health*, 2000, **1**, 156.

[6] M. Wackernagel and W. Rees, *Our Ecological Footprint: Reducing Human Impact on the Earth*, New Society Publishers, 1996.

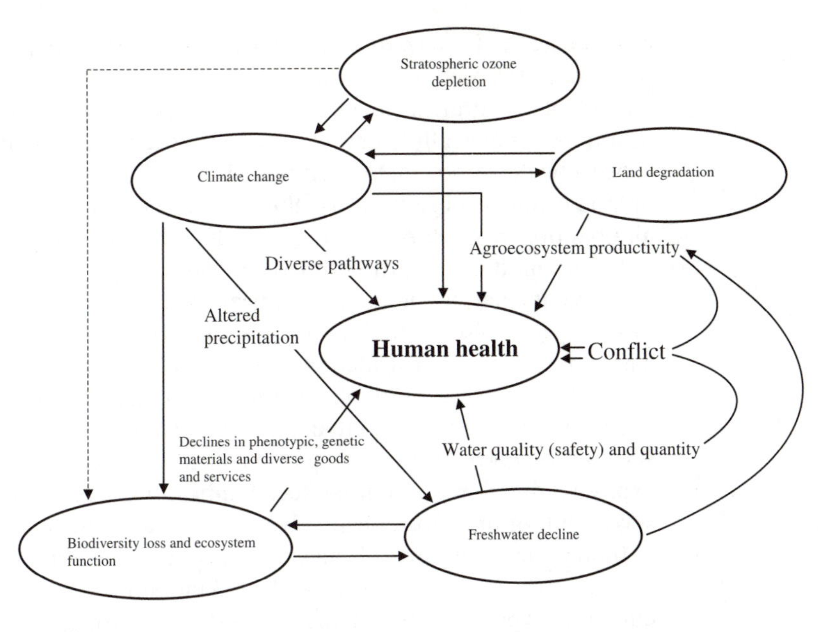

Figure 1 Complex interactions between the biophysical environment and human health

the combustion of fossil fuels causes local air pollution, regional acid rain, and contributes to the global accumulation of carbon dioxide in the lower atmosphere. There is a complex interaction between humans and the environment, with consequences for human health (see Figure 1). These interactions are now significantly affecting nature's goods and services at all scales.

Biodiversity is being reduced at unprecedented speed; there is depletion of non-replenishable resources and alteration of local, regional and global climatic conditions.[7] In recent decades there has been a growing recognition of these changes, and in particular the 'global environmental changes' of damage to the stratospheric ozone layer by the emission of ozone-destroying gaseous emissions (especially chlorofluorocarbons – CFCs), and the accumulation of heat-trapping greenhouse gases (GHG) in the lower atmosphere. These changes entail a range of hazards to human population health, where the future impacts on health are uncertain but predictable in general terms.[8]

Here, we review these changes and their health impacts at the local, regional and global scales. We begin with the familiar, mostly local, topics of local air pollution and microbially contaminated water. We subsequently move to the larger-scale, more contemporary issues of global environmental changes and their associated risks to population health and to social-political security.

Air Pollution, Global and Regional – Respiratory and Other Adverse Health Effects

Anthropogenic activities, particularly the burning of fossil fuels, have degraded the quality of both indoor and outdoor air, with adverse health effects at all

[7] F. S. Chapin III, E. S. Zavaleta, V. T. Eviner, R. L. Naylor, P. M. Vitousek, H. L. Reynolds, D. U. Hooper, S. Lavorel, O. E. Sala, S. E. Hobbie, M. C. Mack and S. Díaz, *Nature*, 2000, **405**, 234.

[8] A. J. McMichael and A. Haines, *Br. Med. J.*, 1997, **315**, 805.

physical scales. This is particularly the case with urban air pollution, which has, in recent decades, become a worldwide public health problem.[9,10] That is, although it does not constitute an integrated 'global' environmental change, its widespread occurrence as a locally generated and locally acting health hazard makes it, now, a 'worldwide' problem. An estimated 130 000 premature deaths and 50–70 million incidents of respiratory illness occur each year due to episodes of urban air pollution in developing countries, half of them in East Asia.[11] An even greater toll of chronic disease is attributable to long-term exposures to urban air pollution.[12]

A recent analysis of three European countries estimated that 6% of total mortality (over 40 000 attributable cases per year) was caused by air pollution.[13] This study also demonstrated that air pollution resulting from motorized traffic accounted for over 16 million person-days of restricted activities. For the world as a whole, it has been estimated that if there were worldwide compliance with the Kyoto Protocol on climate change, the concomitant reduction in particulate-matter exposure, due to reduced fossil fuel combustion, would result in approximately seven million premature deaths being avoided by 2020.[14]

Indoor air pollution is also a cause of concern, especially where biomass fuels or coal, are used for heating and cooking, and where ventilation, including chimneys, is poor.[15] About half the world's population still relies on unprocessed solid fuels (coal and biomass – wood, animal dung and crop residues) for household cooking and heating, often producing pollution that is orders of magnitude more health-damaging per unit of energy than liquid or gaseous fuels.[16] Rough estimates by WHO indicate that, worldwide, as many as 2.5 million people die prematurely each year from exposure to the combustion products of solid household fuels.[17] These estimates indicate that dirty household air has a similar magnitude of health impact to that attributed to contaminated water at the household level, that is, about 6–7% of the global burden of disease – considerably more than the 0.5% attributed to urban ambient air pollution in some widely-cited estimates.

Microbial Water Pollution and Sanitation

Water plays an important role in many of the Earth's systems. Covering over 70% of the planet's surface, oceanic waters play an important stabilization role in

[9] World Bank, *World Development Report. Development and the Environment*, Oxford University Press, Oxford, 1992.

[10] World Resources Institute, *World Resources 1998–99: A Guide to the Global Environment; Environmental Change and Human Health*, Oxford University Press, Oxford, 1998.

[11] D. Maddison, *A Meta-analysis of Air Pollution Epidemiological Studies*, Centre for Social and Economic Research on the Global Environment, University College, London, 1997.

[12] A. J. McMichael, H. R. Anderson, B. Brunekreef and A. Cohen, *Int. J. Epidemiol.*, 1998, **27**, 450.

[13] N. Kunzli, R. Kaiser, S. Medina, M. Studnicka, O. Chanel, P. Filliger, M. Herry, F. Horak, V. Puybonnieux-Texier, P. Quenel, J. Schneider, R. Seethaler, J. C. Vergnaud and H. Sommer, *Lancet*, 2000, **356**, 795.

[14] Working Group on Public Health and Fossil Fuel Combustion, *Lancet*, 1997, **350**, 1341.

[15] N. Bruce, R. P. Padilla and R. Albalak, *Bull. World Health Org.*, 2000, **78**, 1078.

[16] A. Reddy and T. Johannsen, *Energy since Rio*, United Nations Development Programme, 1997.

[17] World Health Organisation, *Health and Environment in Sustainable Development*, Document WHO/EHG/97.8, WHO, Geneva, 1997.

the global climate system, sustain an enormous diversity of plants and animals, from minute phytoplankton to whales[18] and act as an important sink for the by-products of many human activities. Riparian and aquifer waters also have a role in regulating the climate system, although their immediate benefit is on a local or regional scale. At these scales they are important providers of water for drinking, irrigation and industry. Riparian waters are also important for the provision of food, and hydroelectric power.

Water is essential for human health, and yet for many of the world's poorer populations one of the greatest environmental threats to health remains a lack of clean water and sanitation. A recent report[19] estimated that more than half the world's population (some three billion people) live in squalor without access to proper sanitation. One billion people have no access to safe water at all, an important factor in the enormous global burden of diarrhoeal illness and early childhood mortality, when approximately 5000 children die every day from water-related illnesses.

Human activities are having adverse effects on these water stocks, which in turn have adverse consequences for human health. Recent warming of the oceans,[20] presumably, at least in part, because of anthropogenic activities, can have adverse consequences. Warmer oceans raise the sea level, affect the global climate system,[21,22] damage corals[23] and alter the distribution of fishstocks[24] and behaviour of marine ecosystems.[25] The overharvesting of marine species is reducing the trophic level of the average marine harvest[26] and many fisheries face collapse.[27]

Population growth, agricultural irrigation and industrial needs are placing ever-greater demands on limited water resources. Approximately 40% of the world's population now face some level of water shortage, and underground water reserves in many countries are being used faster than they are replenished. On every continent, water tables are dropping – under the north China plain, which produces nearly 40% of the Chinese grain harvest, the fall averages 1.5 m a year. Regions under the most water pressure include China's Yellow River basin, the Middle East and the Aral Sea region of Central Asia.[28–32] Most of the water from these sources is used for irrigation and industry rather than drinking. Unless

[18] World Commission on Environment and Development, *Our Common Future*, Oxford University Press, Oxford, 1987.

[19] World Water Commission, *A Water Secure World. Visionfor Water, Life and the Environment*, 2000.

[20] S. Levitus, J.I. Antonov, T.P. Boyer and C. Stephens, *Science*, 2000, **287**, 2225.

[21] S.-Y. Hong and E. Kalnay, *Nature*, 2000, **408**, 842.

[22] R.H. Zhang, L.M. Rothstein and A.J. Busalacchi, *Nature*, 1998, **391**, 879.

[23] P. Pockley, *Nature*, 1999, **400**, 98.

[24] C.M. O'Brien, C.J. Fox, B. Planque and J. Casey, *Nature*, 2000, **404**, 142.

[25] E. Sanford, *Science*, 1999, **283**, 2095.

[26] D. Pauly, V. Christensen, J. Dalsgaard, R. Froese and F. Torres Jr., *Science*, 1998, **279**, 860.

[27] J.A. Hutchings, *Nature*, 2000, **406**, 882.

[28] L.R. Brown, *Worldwatch Issue Alert*, 2000, **1**.

[29] P.H. Gleick, *Sci. Am.*, 2001, February, 28.

[30] T.F. Homer-Dixon, *Environment, Scarcity and Violence*, Princeton University Press, 1999.

[31] J.R. McNeill, in *Something New Under the Sun. An Environmental History of the Twentieth-century World*, New York, 2000.

[32] S. Postel, *Sci. Am.*, 2001, February, 34.

changes are made, it is estimated that within the next two decades the use of water by humans will increase by about 40% and demand will outstrip available supplies.

Water-related political and public health crises loom in several regions within decades, including the Middle East, northern Africa and parts of South Asia. India, which had a supply of 5500 m^3 per person per year in 1950, currently has around 1800 m^3 per person (close to the recognized minimum requirement), and this will fall by a further quarter over the coming 25 years.[33] Environmental conflict and security over scarce resources has implications for human health, and this is discussed further in Section 3.

Chemical Contamination of Air, Water, Food and Soil

Soil plays an important part in the regulation of the climatic system, and provides food, which is essential for human health. The existence of a sufficient global stock of fertile soil is essential to ensure adequate food production for an indefinite period. However, soil is a finite resource and has been depleted on a global scale.[34] In fact, we have entered the 21st century with an estimated one-third of the world's productive land moderately or severely damaged, by erosion, compaction, salination and waterlogging. Soil has also been damaged by chemicalization. For instance, the use of chemicals has reduced the organic content of soils[10,35,36] and resulted in contamination with heavy metals and persistent pollutants.[37]

Chemicals are in common use in many industrial and manufacturing processes and have provided numerous benefits to society by, for example, ensuring food security and protecting health.[38] Many of these chemicals have accumulated in air, water and soil. This chemicalization of soil and waterways is likely to increase as the use of nitrogenous fertilizer increases, particularly in Latin America and Africa. Already the past half-century's combination of huge increases in nitrogenous fertilizer use, in livestock production, and in the combustion of fossil fuels, has added greatly to the level of biologically active ('fixed') nitrogen within the biosphere. Humankind now produces more fixed nitrogen annually than do the world's natural processes (vulcanism, lightning, naturally-occurring rhizomes, *etc.*).[39] This has contributed to the acidification of soils and has resulted in increasingly high nitrate levels in ground water. In China, for example, nitrate levels are already well above the WHO standard, set in relation to public health risks, and these may well double over the coming half-century.[40]

Exposure to these chemicals is also proving to have detrimental consequences

[33] R. Cassen and P. Visario, *Br. Med. J.*, 1999, **319**, 995.

[34] D. Pimentel, *Ecosystem Health*, 2000, **6**, 221.

[35] D. J. Greenland, P. J. Gregory and P. H. Nye, in *Feeding a World Population of More than Eight Billion*, ed. J. C. Waterlow, D. G. Armstrong, L. Fowden and R. Riley, Oxford University Press, 1998.

[36] C. Lok, *Nature*, 2001, **409**, 969.

[37] European Environment Agency and United Nations Environment Programme, *Down to Earth: Soil Degradation and Sustainable Development in Europe. A Challenge for the 21*st Century, http://themes.eea.eu.int/binary/e/envissue16.pdf (27.12.00), 2000.

[38] B. E. Fisher, *Environ. Health Perspect.*, 1999, **107**, A18.

[39] P. M. Vitousek, H. A. Mooney, J. Lubchenco, and J. M. Melillo, *Science*, 1997, **277**, 494.

[40] K. Strzepek, EMF Conference, *Snowmass*, Colorado, 2000.

for human health.[41,42] In particular, persistent organic pollutants (POPs) and heavy metals impact on human health, and as such are two groups of hazardous chemicals that have received special attention in recent years.

POPs are of particular concern as they have four common properties: high toxicity, persistence, a special affinity for fat, and high mobility. POPs comprise semi-volatile compounds, which may occur in solid, liquid or gaseous form, depending on the temperature. They have a propensity to evaporate and travel long distances (often thousands of miles in air, water currents, and through the food web; so much so that they are of global concern),[42] and become more concentrated higher in the food chain and with time. As POPs have an affinity for fat, there are major intergenerational human health concerns, as the chemicals can be passed from mother to child during breastfeeding. This transmission can disrupt development of the child and lead to long-term damage. Whilst the effects of POPs on human health are unclear, the weight of evidence suggests that high levels of long-term exposure may lead to birth defects, infertility, increased susceptibility to disease, diminished IQ and an increase in cancers.

The high mobility of POPs can be seen in the evidence from the Arctic Region. In the 1950s a thick haze observed in the North American Arctic was traced to sulfates and dust particles that originated in lower latitudes. This long-range mobility is an example of regional environmental change involving 'transboundary pollution'. In 1972 the Stockholm Declaration on the Human Environment declared that states have a responsibility 'to ensure that activities within their jurisdiction or control do not damage the environment of other states',[43] and this became the foundation for establishing the Long-Range Transboundary Air Pollution (LRTAP) regime.

The bioaccumulation of these chemicals occurs when organisms at the base of terrestrial and marine food chains – for example lichen or phytoplankton – absorb the pollutants. These then work their way through the food chain and are eventually consumed by the local human population. The affinity for fat that is displayed by POPs is of particular concern for the populations of these more northern regions, particularly as the same fat which allows them to survive the temperatures at these latitudes is also a store for these pollutants.[44] The United Nations Environment Program (UNEP) responded to international concerns over these persistent chemicals and mounted a campaign for an international ban. This campaign has focused on the twelve most dangerous chemicals – the so-called 'dirty dozen' (see Table 1) – and the UNEP has recently drafted a treaty to ban the production and use of these chemicals.

Heavy metals (for example, mercury, lead and gold) are also of great concern for human health, and have been involved in numerous health disasters. A classic example is the case of Minamata disease. In 1956, hundreds of people in the Japanese coastal town of Minamata were seriously affected by methylmercury

[41] United Nations Environment Program, *Global Environment Outlook 2000*, Earthscan, London, 1999.

[42] World Wildlife Fund, *Issue Brief – Persistent Organic Pollutants: Hand-me-down Poisons that Threaten Wildlife and People*, World Wildlife Fund: http://www.worldwildlife.org/toxics/progareas/pop/pop.pdf, Washington DC, 1999.

[43] O. R. Young, *Environment*, 1999, **41**, 20.

[44] D. J. Tenenbaum, *Environ. Health Perspect.*, 1998, **106**, A64.

Table 1 The 12 persistent organic pollutants designated for international action

I. Pesticides	
Hexachlorobenzene (HCB): Fungicide used for seed treatment of wheat, onions, and sorghum. Also found as an industrial by-product	Endrin: Insecticide used mainly on field crops such as cotton and grains
Mirex: Used as a fire retardant in plastics, rubber, and electrical goods	Toxaphene: A mixture of more than 670 chemicals used as an insecticide, primarily to control insect pests on cotton and other crops
Chlordane: Broad spectrum contact insecticide used on agricultural crops	Heptachlor: Insecticide used primarily against soil insects and termites
DDT: Insecticide used on agricultural crops, especially cotton. Currently used primarily for disease vector control	Aldrin and Dieldrin: insecticides used for crops like potatoes, corn and cotton
Note: Contact with the above list can be by breathing contaminated air, eating contaminated food, or by drinking or washing in contaminated water	
II Industrial chemicals	
Polychlorinated biphenyls (PCBs): Used for a variety of industrial applications, including in electrical transformers, as paint additives, and in plastics	Hexachlorobenzene (HCB): Industrial chemical used to make ammunition and synthetic rubber. Also a by-product from the manufacture of other industrial chemicals
Note: PCBs have a documented history of adverse effects in acutely exposed human populations. Human foetal exposures have been associated with neural and development changes, and long-term intellectual function. HCB is toxic *via* inhalation, can affect the nervous system and cause reproductive and developmental defects.	
III Unintentional by-products	
Dioxins: Not produced commercially by intention and have no known use. They are by-products from the production of other chemicals	Furans: A major contaminant from PCBs. By-product often bonded to dioxin
Note: Dioxins and furans can be created in emissions from the incineration of hospital waste, municipal waste, hazardous waste, and car emissions. Toxic effects of dioxins appear to be due to interference with fundamental biochemical messenger systems, including reproductive disturbances, diminished intellectual capacity and cross-generational toxic effects.	

Source: adapted from ref. 42.

poisoning – which affects the nervous system, and has symptoms ranging from slight numbness of fingers to loss of ability to talk and walk – and many victims died.[45] Methylmercury had been discharged from a local chemical production factory and eventually entered the food chain of fish in a local bay.

Around the same time as the Minamata incident there was a similar outbreak of Itai-itai disease, a form of chronic cadmium poisoning that developed in farmers in Toyama, Japan. In this case cadmium-contaminated rice and drinking water was the cause. The cadmium from a mining area and a lead/zinc ore concentration plant reached the affected community *via* a river that they used to irrigate their paddy.

Lead is the most abundant heavy metal and the natural (preindustrial) blood lead concentration of humans is estimated to be much lower than the lowest reported levels in contemporary humans living in remote regions.[45] Although several high-income countries have legislated for new lower standards on environmental lead levels, exposure is still a particular problem in urban environments of low-income countries.[46,47] Childhood lead poisoning (a particular hazard for the neuro-cognitive development of children) is an increasing problem, and high blood lead levels have been widely observed in cities such as Bangkok, Jakarta, Taipei, Santiago and Mexico City.[9,47]

Stratospheric Ozone Layer

An intact stratospheric ozone layer (SOL) prevents excessive health-damaging ultraviolet radiation reaching the Earth's surface. Until the early 1970s this layer was intact (although there was a natural variation in the thickness of the SOL, depending on latitude, season and volcanic eruptions). The depletion of stratospheric ozone, first observed in the 1970s over Antarctica,[48] is primarily caused by the build-up of human-made ozone-destroying gases in the stratosphere, such as the chlorofluorocarbons (CFCs), and is causing an increase in ultraviolet irradiation (UVR) at the Earth's surface. Ambient terrestrial levels of UVR are estimated to have increased by over 10% at mid-to-high latitudes since 1980.[49]

In terms of the adverse effects on human health we can expect, for the first half of the 21st century, an increase in the severity of sunburn and the incidence of skin cancers in fair-skinned populations, and the incidence of various disorders of the eye (especially cataracts). Since cataract accounts for a majority of the tens of millions of cases of blindness in the world, even a marginal impact of increased UVR exposure on their occurrence would be significant. Some UVR-induced suppression of immune functioning may also result, thus increasing susceptibility to infectious diseases and perhaps reducing vaccination efficacy.[50,51]

[45] A. J. McMichael, T. Kjellstrom and K. Smith, in *International Public Health*, ed. M. Merson, R. Black, and A. Mills, Aspen Press, Gaithersburg, MD, 2000.

[46] L. M. Schell, in *Effects of Pollutants on Human Prenatal and Postnatal Growth: Noise, Lead, Polychlorobiphenyl Compounds and Toxic Wastes, Yearbook of Physical Anthropology*, Vol. 34, 1991, pp. 157–188.

[47] S. Tong, Y. E. von Schirnding and T. Prapamontol, *Bull. World Health Org.*, 2000, **78**, 1068.

[48] J. C. Farman, B. G. Gardiner and J. D. Shanklin, *Nature*, 1985, **315**, 207.

[49] R. McKenzie, B. Connor and G. Bodeker, *Science*, 1999, **285**, 1709.

[50] United Nations Environment Program, *Environmental Effects of Ozone Depletion*, UNEP, Lausanne, Switzerland, 1998.

[51] World Health Organisation, *Ultraviolet Radiation*, Environmental Health Criteria, No. 160, WHO, Geneva, 1994.

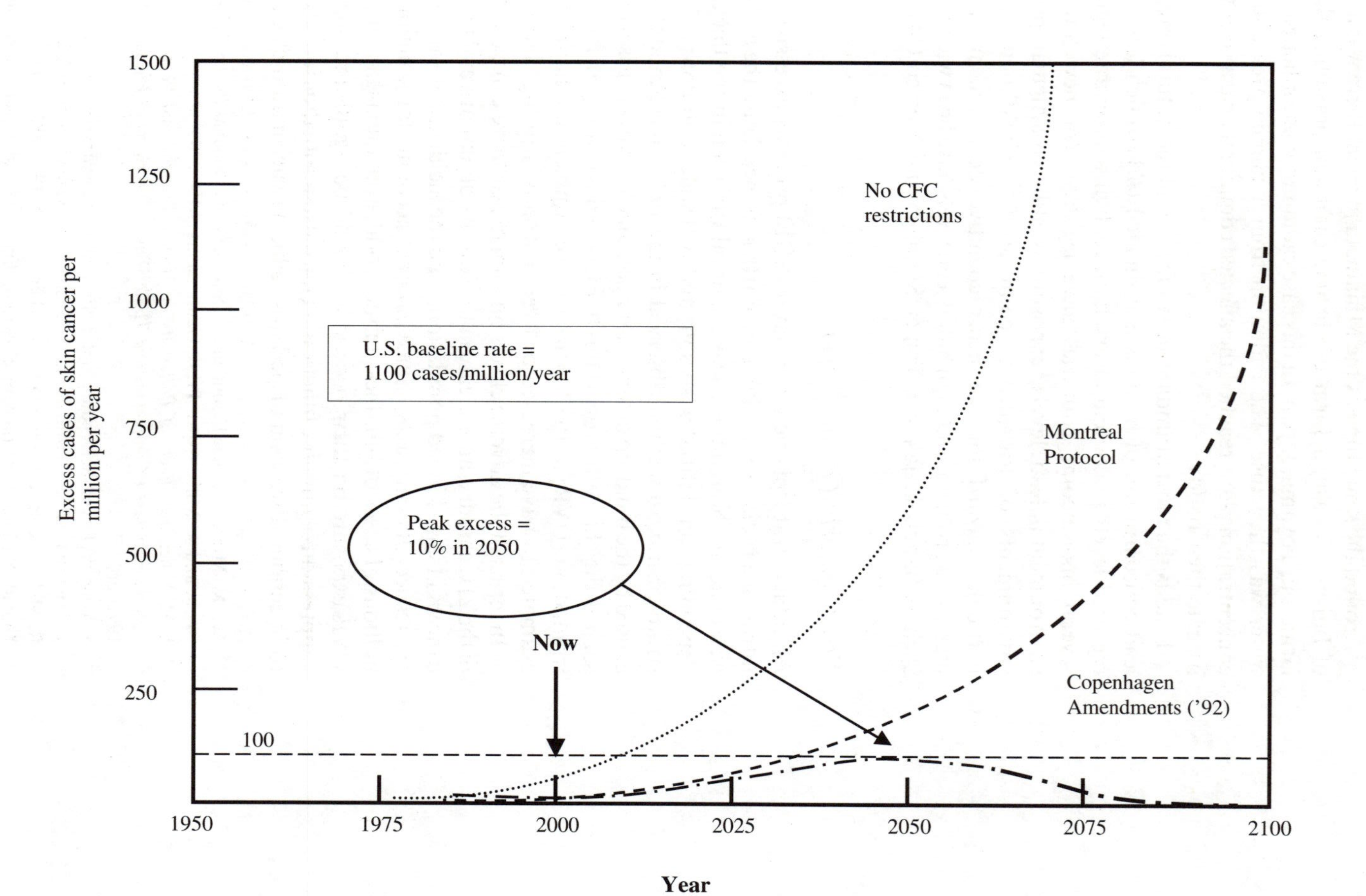

Figure 2 Estimates of ozone depletion and skin cancer incidence to examine the Vienna Convention achievements (adapted from ref. 53)

On the other hand, recent work suggests that increased UVR at high latitudes could have some population health benefits, helping to prevent certain autoimmune diseases.[52] Further research is needed to quantify the optimum dose of sunlight exposure, depending on factors including age, ethnicity, behaviour, latitude, cloud cover and stratospheric ozone depletion (SOD). The health impacts of SOD are difficult to quantify, but broad projections of the burdens of skin cancer and other UV-related disease under various scenarios of emission containment have been estimated[53,54] (see Figure 2).

A potentially more important, although much more indirect, health detriment could arise from ultraviolet-induced impairment of photosynthesis on land (terrestrial plants) and at sea (phytoplankton).[55] Such an effect could both reduce the world's food production, and also harm the oceanic carbon sink.[56] However, few quantitative data are yet available.

Biodiversity – Destruction and Invasion

Through humankind's reproductive and technological 'success', we have occupied, damaged or eliminated the natural habitat of many other species. Biologists estimate that this mass extinction may cause around one third of all species that were alive in the 19th century to be gone before the end of the 21st century.[7,57] The loss of various key species weakens ecosystems, with many potential adverse consequences for humans as 'nature's goods and services' decline.[1,58] Examples include disturbing the ecology of vector-borne infections – for example, by reducing or eliminating mosquito predators – thereby enhancing transmission of infections such as yellow fever and malaria. Other examples include damage to food-producing systems that depend on pollinators and the predation of pests, and impairing the cleansing of water and the circulation of nutrients that normally pass through ecosystems.

We would also lose a rich repertoire of genetic and phenotypic material. To maintain the hybrid vigour and environmental resilience of 'food' species, a diversity of wild species needs to be preserved as a source of genetic additives. Similarly, a high proportion of modern medicinal drugs in western medicine has natural origins, and many defy synthesis in the laboratory. Scientists test thousands of novel natural chemicals each year, seeking new drugs to treat HIV, malaria, drug-resistant tuberculosis, cancers and so on.[59]

The other side of this coin is the accelerating spread of 'invasive' species, as long-distance trade, tourism and migration increase in intensity.[60] For example,

[52] A.-L. Ponsonby, A. J. McMichael and I. van der Mei, *Toxicology*, 2001, in press.

[53] H. Slaper, G. J. M. Velders, J. S. Daniel, F. R. de Gruijl and J. C. van der Leun, *Nature*, 1996, **384**, 256.

[54] J. Longstreth, F. R. de Gruijl, M. L. Kripke, S. Abseck, F. Arnold, H. Slaper, G. J. M. Velders, Y. Takiwaza and J. C. van der Leun, *J. Photochem. Photobiol. B: Biol.*, 1988, **46**, 20.

[55] P. J. Neale, R. F. Davis and J. J. Cullen, *Nature*, 1998, **392**, 585.

[56] C. B. Field, M. J. Behrenfeld, J. T. Randerson and P. Falkowski, *Science*, 1998, **281**, 237.

[57] S. L. Pimm, G. J. Russell, J. L. Gittleman and T. M. Brooks, *Science*, 1995, **269**, 347.

[58] G. C. Daily, in *Nature's Services: Societal Dependence on Natural Ecosystems*, Washington DC, 1997.

[59] P. A. Cox, *Science*, 2000, **287**, 44.

[60] T. Low, *Feral Future: The Untold Story of Australia's Exotic Invaders*, Viking, 1999.

the spread of water hyacinth in eastern Africa's Lake Victoria, introduced from Brazil as a decorative plant, has provided a micro-environment for the proliferation of diarrhoeal disease bacteria and water snails that transmit schistosomiasis.[61]

Climate Change

The greenhouse effect is a natural phenomenon, ensuring relative stability in aggregate global temperatures. However, increasing fossil fuel combustion, especially in high-income industrialized societies over recent decades, along with continuing and widespread forest clearance, has augmented the greenhouse effect in the lower atmosphere. This is a clear manifestation of the unprecedented large-scale environmental changes that humankind is now inducing. Global climate change poses substantial risks to human health over the coming century, and arresting this process presents a major challenge to the world community.

Global temperatures have risen over the last one a half centuries, and the particular pattern of warming suggests that human-induced alteration of atmospheric composition, mainly from burning fossil fuels, has been an increasingly prominent cause.[62] Average world temperatures increased over the last quarter of the 20th century, mostly due to human influence, and weather patterns in many regions displayed increasing instability in the 1990s.[63] Current predictions are for a further rise of around 2–3 °C (although one report[64] suggests a rise of 5.5 °C) within a plausible range of 1.4–5.8 °C[65] over the next century. This represents a greater rate of change in temperature than at any time in the last 100 000 years, except perhaps at some stages during the sometimes-hectic transition to the Holocene 10–15 000 years ago.

Climate change would lead to health effects that would encompass *direct* and *indirect*, *immediate* and *delayed effects*[8] (see Table 2). Some health outcomes in some populations would be beneficial. For example, some tropical regions may become too hot for mosquitoes, and winter cold-snaps would become milder in temperate-zone countries where death rates typically peak in winter time – but most of the anticipated health effects would be adverse.[62]

Direct health effects would include changes in mortality and morbidity from an altered pattern of exposure to thermal extremes, the respiratory health consequences of increased exposures to photochemical pollutants and aeroallergens, and the physical hazards of the increased occurrence, in at least some regions, of storms, floods or droughts. Intensified rainfall, with flooding, can overwhelm urban wastewater and sewer systems, leading to contamination of drinking water supplies, and would be most likely to occur in large crowded cities where

[61] P. R. Epstein, H. F. Diaz, S. Elias, G. Grabherr, N. E. Graham, W. J. M. Martens, A. Fenwick, A. K. Chessmond and M. A. Amin, *Bull. World Health Org.*, 1998, **59**, 777.

[62] Intergovernmental Panel on Climate Change, *Second Assessment Report. Climate Change 1995* (Vols. I, II, III), Cambridge University Press, Cambridge, 1996.

[63] D. R. Easterling, G. A. Meehl, C. Parmesan, S. A. Changnon, T. R. Karl and L. O. Mearns, *Science*, 2000, **289**, 2068.

[64] P. M. Cox, R. Betts, C. D. Jones, S. A. Spall and I. J. Totterdell, *Nature*, 2000, **408**, 184.

[65] Intergovernmental Panel on Climate Change, *Third Assessment Report*, Cambridge University Press, Cambridge, 2001.

Table 2 Mediating processes and direct and indirect potential effects on health of changes in temperature and weather

Mediating process	*Health outcome*
Direct effects	
Exposure to thermal extremes	Changed rates of illness and death related to heat and cold
Change frequency or intensity of other extreme weather events	Deaths, injuries, psychological disorders; damage to public health infrastructure
Indirect effects	
Disturbances of ecological systems:	
Effect on range and activity of vectors and infective parasites	Changes in geographical ranges and incidence of vector-borne disease
Changed local ecology of water-borne and food-borne infective agents	Changed incidence of diarrhoeal and other infectious diseases
Changed food productivity (especially crops) through changes in climate and associated pests and diseases	Malnutrition and hunger, and consequent impairment of child growth and development
Sea level rise with population displacement and damage to infrastructure	Increased risk of infectious disease, psychological disorders
Biological impact of air pollution changes (including pollens and spores)	Asthma and allergies; other acute and chronic respiratory disorders and deaths
Social, economic, and demographic dislocation through effects on economy, infrastructure and resource supply	Wide range of public health consequences: mental health and nutritional impairment, infectious diseases, civil strife

Source: ref. 8.

infrastructure is old or inadequate.

Indirect health effects would include alterations in the geographical range (latitude and altitude) and seasonality of certain vector-borne infectious diseases such as malaria, dengue fever, schistosomiasis, leishmaniasis, and Lyme disease. Other examples would include various forms of tick-borne viral encephalitis.[61,66,67]

Many bacteria and protozoa are sensitive to temperature; hence, climate change would also influence various directly transmitted infections, especially those due to contamination of drinking water and food. This is likely to include an influence on seasonal summer-time peaks of food-borne infections, such as salmonellosis. Changes in the pattern of rainfall can disrupt surface water configuration and drinking water supplies. Hence, the occurrence of infectious diseases such as cryptosporidiosis and giardiasis, spread *via* contaminated drinking water, would be influenced by a change in climatic conditions. The

[66] W. J. M. Martens, R. S. Kovats, S. Nijhof, P. de Vries, M. T. J. Livermore, D. Bradley, J. Cox and A. J. McMichael, *Global Environ. Change*, 1999, **9**, S89.
[67] J. A. Patz, *J. Am. Med. Assoc.*, 1996, **275**, 217.

consequences will vary from region to region, but the burdens are likely to fall disproportionately on the poorer populations in low-income countries, and on those made vulnerable by age or pre-existing illness.

Sea-level rise is another environmental consequence of climate change. Oceans are thermally expanding and most glaciers are already shrinking in a warmer world.[68] In consequence, sea level is forecast to rise by approximately 40 cm by 2100.[62] This rate of rise would be several times faster than has occurred over the past century. This is important, since over half of the world's population now lives within 60 km of the sea, and such a rise in sea level could have widespread impacts on public health, especially in vulnerable populations (*e.g.* small island states, coastal Bangladesh and the Nile Delta). A half-metre rise (at today's population) would approximately double the number who experience flooding annually from around 50 million to 100 million. A range of adverse physical and psychological health consequences would result from population displacement and economic disruption due to rising sea levels, agroecosystem decline and freshwater shortages.

Sea-level rise would damage coastal structures and arable land; rising seas would salinate coastal freshwater aquifers, particularly those beneath small islands. Sea-level rise would also affect sewage and wastewater disposal; and would influence the local ecology of certain infectious diseases such as malaria and cholera. Indeed, for cholera it is becoming evident that there is a rich and complex set of ecological circumstances that influence transmission probabilities. Research findings suggest, increasingly, that the spread of cholera is facilitated by warmer coastal and estuarine waters and their associated algal blooms.[69] This means we must add an ecological dimension to the traditional transmission model of cholera: person-to-person spread *via* the direct faecal contamination of local drinking water is not the entire explanation.

The prospect of climate change and the anticipated range of impacts on food production, population health and social well being add a new dimension of urgency to our need to find ways of living sustainably. With respect to food production, the goal is to improve crop yields while leaving the natural resource base intact. The main ways in which climate change would affect terrestrial food production are listed in Table 3.

Scientists have used dynamic crop growth models to simulate the effects of climate change, in conjunction with increased atmospheric carbon dioxide, on cereal crop yields (which represent almost two-thirds of world food energy). One pioneering modelling study estimated the additional number of hungry people attributable to standard projections of climate change by the year 2060, within a range of plausible future trajectories of demographic, economic and trade-liberalization processes.[70] The estimate varied, depending on the mix of other assumptions, between an additional 40 million and 300 million relative to a future background total of around 600 million hungry people.

Regionally, most of this nutritional adversity would occur in sub-Saharan

[68] L. G. Thompson, T. Yao, E. Mosley-Thompson, M. E. Davis, K. A. Henderson and P.-N. Lin, *Science*, 2000, **289**, 1916.

[69] M. Pascual, X. Rodó, S. P. Ellner, R. R. Colwell and M. J. Bouma, *Science*, 2000, **289**, 1766.

[70] M. Parry, C. Rosenzweig, A. Iglesias, G. Fischer, and M. T. J. Livermore, *Global Environ. Change*, 1999, **9**, S51.

Table 3 How might climate change affect terrestrial food yields?

1. Temperature effects on plant physiology
2. Soil moisture effects on plant physiology
3. Carbon dioxide fertilization effects: gains in plant water-use efficiency
4. Climatic influences on plant disease occurrence
5. Climatic influences on crop losses *via* pest species
6. Damage due to extreme weather events: floods, droughts, *etc.*
7. Sea-level rise: salination and inundation of coastal land

Africa. The resultant additional hunger and malnutrition would increase the risk of infant and child mortality and cause physical and intellectual stunting. In adults, energy levels, work capacity and health status would be compromised. The uncertainties inherent in this sort of attempt to model future climate change impacts on world food production are well illustrated by the spread of estimates obtained in other global studies.[71,72]

It is important to note also the potential impacts of climate change upon food yields from the marine and freshwater aquatic environment (approximately one-sixth of all protein consumed by the world population is of aquatic origin, and in many developing countries it accounts for the majority of animal protein). The Intergovernmental Panel on Climate Change (IPCC) in its Third Assessment Report has noted that, while weather impacts and seasonal rhythms have long been recognized by the global fishing industry, decadal-scale shifts in climate have only recently been acknowledged as a factor in fish and marine ecosystem dynamics.[65] In fact, various life-stages of fish populations are sensitive to temperature: spawning, growth rates (in part because of temperature influences on food availability), migratory patterns and breeding routes.

The important question about how global climate change is likely to affect food production remains complex and riven with uncertainties. There are finite, and increasingly evident, limits to agroecosystems and to wild fisheries. Our capacity to maintain food supplies for an increasingly large and increasingly expectant world population will depend on maximizing the efficiency and sustainability of production methods, incorporating socially beneficial genetic biotechnologies, and taking pre-emptive action to minimize the future course of detrimental, ecologically damaging, global environmental changes.

Although there is currently no basis for making overall estimates of the direct costs to society of the health impacts of climate change, an attempt has been made to provide some guidance.[65] This catalogued some of the recent approximate estimates that have been published of the impacts on national economies of major infectious disease outbreaks, such as might occur more often under conditions of climate change, and included the following examples:

- Outbreak of plague-like disease in Surat, Northwest India in 1994 cost an estimated US$ 3 billion in lost revenues to India alone.

[71] C. Rosenzweig, A. Iglesias, X. B. Yang, P. R. Epstein and E. Chivian, *Climate Change and U.S. Agriculture: The impacts of warming and extreme weather events on productivity, plant diseases and pests*, Center for Health and the Global Environment, Harvard Medical School, USA, 2000.

[72] P. Winters, R. Murgai, A. de Janvry, E. Sadoulet and G. Frisvold, in *Global Environmental Change and Agriculture*, ed. G. Frisvold and B. Kuhn, Cheltenham, Gloucester, 1999.

- Cost of the 1994 dengue haemorrhagic fever (DHF) epidemic in Thailand was estimated to be US$ 19–51 million.
- Cost of the 1994 epidemic of dengue/DHF in Puerto Rico was estimated to be US$ 12 million for direct hospitalization costs alone.

In a world where anthropogenic activities are having an impact on Earth's life-supporting systems, we must consider how pressures on these systems have the potential to lead to environmental conflict and insecurity, with all the adverse effects on human health. This is the focus of the next section.

3 Environmental Conflict and Security

As human numbers expanded over the millennia, there was increased exploitation of nature's capital, and increasing territorial expansion. Territorial expansion, a desire for greater wealth, and hegemonic power, often led to conflict between rival groups. Although such conflict at different tempero-spatial scales is multifactorial, complex and contentious, the competition for natural resources is a key factor.[73,74] The World Commission on Environment and Development argued that nations have often fought to assert or resist control over raw materials, energy supplies and land.[18] One might therefore envision that the risk of conflict will significantly increase in the near future because of increased natural resource scarcity, much of it consequent to declining environmental goods and services.

The Persian Gulf War of 1991 is a recent example of major conflict triggered by concern over an environmental resource: oil.[75] Other examples from recent, lesser known, resource-associated conflicts include those from India, the Philippines and the West African states of Mauritania and Senegal.[76]

In many countries, agricultural production is increasingly dependent on irrigation, and this is likely to lead to conflict where there are existing tensions over access to freshwater supplies.[77] Many river systems (and thus scarce water resources) are shared uneasily between neighbours in unstable regions: the Nile, the Ganges, the Mekong, the Jordan and the Tigris and Euphrates rivers.[75,77] 'Water wars' have therefore been postulated as increasingly likely in future, as population pressures and demands increase, including in the Middle East, and between Ethiopia and Egypt, Lesotho and South Africa, and India and Bangladesh.[75,78]

Motivation and opportunity are two other key factors for the emergence of conflict.[79] Excessive inequality may sow the seeds of resentment and future

[73] T. F. Homer-Dixon and J. Blitt, in *Ecoviolence. Links among Environment, Population and Security*, ed. T. F. Homer-Dixon and J. Blitt, Lanham, MD, 1999.

[74] G. D. Snooks, *The Dynamic Society. Exploring the Sources of Global Change*, Routledge, 1996.

[75] T. F. Homer-Dixon, *Int. Security*, 1994, **19**, 5.

[76] T. F. Homer-Dixon, J. H. Boutwel and G. W. Rathjens, *Sci. Am.*, 1993, **268**, 16.

[77] P. H. Gleick, *The World's Water. The Biennial Report on Freshwater Resources 1998–1999*, Island Press, Washington DC, 1998.

[78] P. H. Gleick, *The World's Water. The Biennial Report on Freshwater Resources*, Island Press, Washington DC, 2000.

[79] V. Percival and T. F. Homer-Dixon, in *Ecoviolence. Links among Environment, Population and Security*, ed. T. F. Homer-Dixon and J. Blitt, Lanham, MD, 1999.

conflict, but conflict is unlikely to occur, paradoxically, until the inequality is reduced.[80] This is because it is irrational for a weaker population to revolt or invade with no chance of victory. In Rwanda, the 1994 genocide was not instigated by the most deprived section of its population.[79]

Even if the 'sustainability transition'[81] accelerates sufficiently to enable a 'factor four' world – doubling resource output yet halving the ecological cost[82,83] – it is still likely that the per capita availability of water, arable land and other critical environmental resources will decline. Spectacular technological improvements in the exploration and recovery of oil[84] have not abolished concerns that the end of cheap oil is likely in this century:[85,86] future oil wars are therefore also possible.

On a more speculative level, there is the possibility that climate change may interact with natural resource stresses – such as water scarcity – and expanding human populations, to increase the possibility of conflict. Many parts of Africa already experience a less than favourable agricultural climate, and under various climate models this already less-than-favourable climate is forecast to deteriorate in the second half of this century.[70] In that case, the likelihood of conflict would increase.

Global warming has also been identified as possibly intensifying the El Niño Southern Oscillation (ENSO).[87] Stronger, more frequent El Niños and La Niñas would be likely to lead to increased adverse social, economic, and health consequences in different regions over a widespread area.[69,88] These, in turn, would tend to increase the risk of conflict in resource-scarce areas, for example by increasing regional food scarcity through intensified droughts.

Ecosystem changes, especially from deforestation, the infilling of wetlands and the replacement of coastal mangroves by aquaculture, may also have adverse effects on resource security, including through an interaction with rising seas and more intense storms. The flooding of the Yangtze river basin in 1998, in China, has been attributed to a complex web of factors, including heavy rain associated with an El Niño event, deforestation which increased water runoff, and more intensive cultivation of lakes and wetlands in the river basin which reduced their 'sponge' function. Changes have been called for in Chinese ecosystem management to try to avert such events in the future.[89]

Falling biodiversity may not obviously appear to aggravate the risk of conflict. The loss of genetic information will reduce the isolation of useful chemicals and the discovery of potentially useful biological principles, but is unlikely to lead to

[80] A. Hurrell, in *Security and Inequality*, ed. A. Hurrell and N. Woods, Oxford University Press, 1999.

[81] A. J. McMichael, K. R. Smith and C. F. Corvalan, *Bull. World Health Org.*, 2000, **78**, 1067.

[82] P. Hawken, A. B. Lovins and L. H. Lovins, *Natural Capitalism: Creating the Next Industrial Revolution*, Little Brown & Company, 1999.

[83] E. von Weizsäcker, A. B. Lovins and L. H. Lovins, *Factor 4. Doubling Wealth – Halving Resource Use. A New Report to the Club of Rome*, Earthscan, London, 1997.

[84] J. Rauch, *Atlantic Monthly*, 2001, **287**, 35.

[85] C. B. Hatfield, *Nature*, 1997, **388**, 618.

[86] H. S. Houthakker, *Nature*, 1997, **388**, 618.

[87] A. Timmermann, J. Oberhuber, A. Bacher, M. Esch, M. Latif and E. Roeckner, *Nature*, 1999, **398**, 694.

[88] M. J. Bouma, R. S. Kovats, S. A. Goubet, J. Cox and A. Haines, *Lancet*, 1997, **350**, 1435.

[89] J. N. Abramovitz, in *Averting Unnatural Disasters*, ed. L. Starke, New York, 2001.

war. However, reduced ecosystem function, consequent to falling and altered biodiversity, may interact with climate change to cause further deforestation and ecosystem collapse, for example by the loss of 'keystone species' or by changing the flowering and fruiting patterns of the tropical rainforest canopy.[90,91]

There are numerous other mechanisms by which damaged environmental 'goods and services' may cause economic harm, and increase the risk of conflict. Several worst-case scenarios could even lead to global conflict. These include runaway global warming, food-scarcity associated nuclear conflict involving South Asia or China and disruption of the North Atlantic climatic 'conveyor belt'. This 'climatological service' effectively subsidizes European civilization by providing significant, free warming.[92] Its loss would disrupt European agriculture and greatly increase European energy needs. It is hard to imagine that the technologically and militarily powerful European states would not act to protect their vital interests. Conflict therefore seems likely.[92–94]

The view that the risk of conflict is likely to increase because of environmental scarcity – in part due to damage to environmental 'goods and services' – is not universally held, however, and remains controversial. This view has been criticized as not only exaggerated, but as being intrinsically harmful as the 'various resource factors, such as access to fuels and ores, have contributed to state capacities to wage war and achieve security from violence'.[95] However, the same author argues that natural disasters have always occurred, but without threatening national security, and 'if everything that causes a decline in human well-being is labelled a security threat, the term loses any analytical usefulness'.[95]

Nevertheless, we argue that a threshold has been crossed in the scale of human-caused capacity to change the regional and global ecosystem and climatological systems, and that this could contribute to conflict on a global scale. Hence, it is in the rational self-interest of humanity, including its most powerful nations, to seek to reduce these risks, including from conflict, of damage to environmental goods and services.

4 Two Issues: Environmental 'Global Public Goods for Health' and Scientific Uncertainty

Defining Global Public Goods for Health

The environmental goods and services described in the earlier parts of this review are, in many respects, quite distinct from the goods and services that are produced and exchanged between individuals, groups and nations on a daily basis. This distinction may be seen within the concept of '*private*' and '*public*' goods. 'Private goods' are generally traded in the marketplace, and through various pricing mechanisms the market dictates the volume of a good that is demanded (by the

[90] L. M. Curran, I. Caniago, G. D. Paoli, D. Astianti, M. Kusneti, M. Leighton, C. E. Nirarita and H. Haeruman, *Science*, 1999, **286**, 2184.
[91] G. Hartshorn and N. Bynum, *Science*, 1999, **286**, 2093.
[92] W. S. Broecker, *Science*, 1997, **278**, 1582.
[93] C. D. Butler, *Australas. Epidemiol.*, 2001, **8**, 13.
[94] H. J. Schellnhuber, *Nature*, 1999, **402**, C19.
[95] D. Deudney, *Bull. At. Sci.*, 1991, **47**.

consumer) and supplied (by the producer). In simple terms an exchange or transaction takes place between producer and consumer. With private goods such exchange is commonplace, and the market is seen as 'the most efficient way of producing private goods'.[96]

'Public goods', such as nature's 'goods and services', are quite different. Unlike private goods, public goods are rarely the sole preserve of a single producer or consumer, and usually do not fit well within the market mechanism (although one could argue that this has changed, for example with the advent of carbon trading under the Kyoto Protocol). Public goods are viewed as existing along a spectrum from *pure* to *impure public goods*.[96] Pure public goods are *non-excludable* – access to the good is open to all – and *non-rivalrous* – the quantity of the good provided cannot be diminished in character.[96] However, in practice, pure public goods are rare; very few goods have unlimited access, and even fewer have an infinite level of supply. In reality, most fall along a spectrum from *pure* to *impure* (that is, they possess the attributes of non-exclusion and non-rivalry to a lesser degree, or possess a combination of them).

Spatial and temporal dimensions are fundamental to any discussion on environmental issues, and this is no different when considering the notion of the *environment as public good for health*.* If the benefits of a public good are to be seen as global in scale – a global public good (GPG) – then the good must not only display non-exclusive and non-rivalrous characteristics, but must also be *quasi-universal*.[97] To fulfil the latter requirement the good must be of benefit in three tempero-spatial and socio-economic dimensions:

- Countries – a GPG must cover more than one geographical region. Otherwise it would be a *regional public good*, and possibly a *club good* (that is, a good with excludable benefits)
- Socio-economic groups – a GPG must be accessible to all socio-economic segments. That is, a GPG must benefit a broad spectrum of the global population
- Generations – a GPG must meet the needs of present generations without jeopardizing those of future generations

One way of visualizing this difficult conceptualization of the environment as a global public good for health is shown in Figure 3.

Dealing with Uncertainty

Environmental public goods represent a paradox, as their attributes are easier to recognize and discuss when they begin to lose their public good status.[98] More

[96] I. Kaul, I. Grunberg and M. A. Stern (eds.), *Global Public Goods: International Co-operation in the 21st Century*, Oxford University Press, 1999.

[97] A. J. McMichael, C. D. Butler and M. J. Ahern in *Providing Global Public Goods: Making Globalisation Work for All*, ed. United Nations Development Program, in press.

[98] S. J. Buck, *Environ. Ethics*, 1985, **7**, 49.

* Environment refers to the 'goods and services' that the biophysical environment provides, and how these are beneficial to human health.

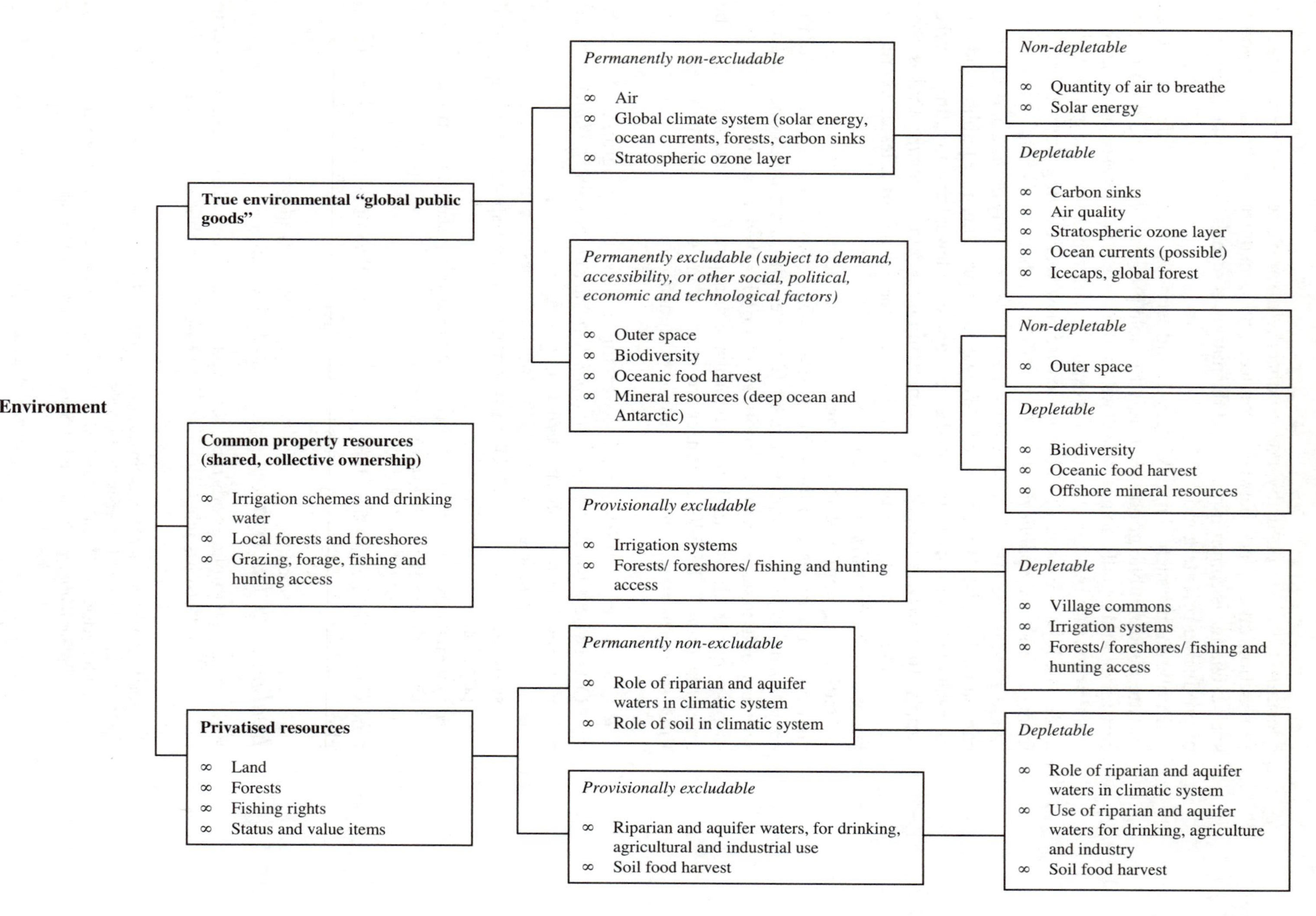

Figure 3 Flow chart of the environment as a *global public good for health* (source: ref. 97)

often than not environmental public goods have been taken for granted. But when these goods begin to deteriorate then civilizations have provided means of protection, at least in terms of local environmental public health goods, in order to limit the 'tragedy of the commons'.[99–100]

Awareness of damage to local environmental public health goods can be traced to the recognition of soil salination by the ancient Sumerians.[101] However, concerns for what we now recognize as environmental global public goods has been much more recent,[102–104] and led to the first United Nations global environment conference, held in Stockholm in 1972. In the past decade there has been an encouraging increase in global public awareness of the importance and fragility of many aspects of the global environment, but with an inability to translate this into effective action.

The scale of human occupation and transformation of the environment is now truly global. It is therefore time to recognize not only the importance of environmental global public goods for health, but also the contribution to be made by the emerging subdiscipline of ecological economics. Unlike classical economics, ecological economics bridges the divide between economics and the natural sciences.[105] It sees the purpose of economics as that of a means rather than an end – that is, as the efficient attainment of socially-agreed goals, rather than as the embodiment and indexation of ongoing compulsive gains in material wealth. These perspectives, if integrated into our scientific and social assessments, will enhance our capacity to deal with the challenges ahead.

These challenges are great, and it is important to emphasize that the complexity of these large-scale environmental issues means that there is inevitably a hierarchy of uncertainties that bear on scientific assessments and policy decision-making. It is for this reason that there is an emerging accommodation with the ideas of the *Precautionary Principle* as a means of decision-making in situations of uncertainty and potentially great future risk. Our review of the human-induced environmental damage in Section 2 shows that all too often we have ignored our environmental heritage and dependency.

The adverse health effects of this ignorance may have been limited, as the environmental damage was on a local scale. However, in today's world our activities are having an impact at a global scale, and global environmental change will become a major theme in public health research, social policy development, and political advocacy in the 21st century. The conceptualization of the environment as a global public good for health should go some way not only in increasing an appreciation for this heritage and dependency, but should also enhance the interface between research and policy. This increased awareness and interface between key stakeholders might lead to effective action to ensure a sustainable future for current and future generations.

[99] P. Dasgupta, *Proc. Brit. Acad.*, 1996, **90**, 165.

[100] G. Gardner, in *State of the World 2001*, ed. L. Starke, New York, 2001.

[101] T. Jacobsen and R. M. Adams, *Science*, 1958, **128**, 1251.

[102] R. Carson, *Silent Spring*, Houghton Mifflin, 1962.

[103] P. R. Ehrlich, *The Population Bomb*, Ballantyne, 1968.

[104] H. A. Mooney and P. R. Ehrlich, in *Nature's Services. Societal Dependence on Natural Ecosystems*, ed. G. C. Daily, 1997.

[105] E. O. Wilson, *Consilience. The Unity of Knowledge*, Alfred A. Knopf, 1998.

5 Acknowledgement

The authors gratefully acknowledge drawing on the ideas of Dr. Colin Butler, Australia National University, from an ongoing collaborative project, for part of this chapter. Figure 3 was produced for the UNDP/WHO collaboration 'Providing Global Public Goods: Making Globalisation Work for All'.

Corporate Environmental Performance

FRANS BERKHOUT

1 Introduction

Environmental performance measurement and reporting are becoming pervasive features of modern business practice. Just as a century ago modern financial accounting procedures began to be developed, so today environmental accounting and management procedures are being created and put in place. There are many reasons for this development. Among them, the desire of managers for better control of environmental aspects of their businesses in order to avoid risks and to capture value, the desire among policymakers and regulators for a richer set of information about company performance to be made widely available, and a demand by many customers and other stakeholders of businesses for better knowledge about how ethically they are behaving. There is a widespread belief that the availability of information will stimulate greater awareness, self-reflection and innovation in business, and reduce the likelihood of poor performance and non-compliance. Information deficits and asymmetries lie at the heart of many environmental problems.

This article presents an overview of the reasons for the growing emphasis on environmental and social transparency in business, under the more general rubric of corporate social responsibility. An assessment is made of current initiatives and previous research related to environmental performance. Finally, some results from a study investigating the environmental performance of European companies are presented. The aim is to provide an introduction to a growing field of research and practice which seems likely to have profound impacts on the environmental behaviour of business.

2 Corporate Social Responsibility

The debate about corporate social responsibility has a long history, and turns on the question of whether firms (or the managers who run them on behalf of shareholders) have responsibilities to society beyond the making of profits.[1] For instance, Milton Friedman, taking a contrary position, argued that: '. . . there is one and only one responsibility of business – to use its resources and engage in

[1] P. Arlow and M. J. Gannon, Social responsiveness, corporate structure and economic performance, *Acad. Manage. Rev.*, 1982, **7**, 235–241.

Issues in Environmental Science and Technology, No. 17
Global Environmental Change

activities designed to increase profits so long as it stays within the rules of the game, which is to say, engages in open and free competition without deception or fraud'.[2] Friedman was responding to an earlier phase of concern about the social and environmental consequences of business activity. He argued that the practical impact of 'social responsibility' must always be to divert management effort away from normal competition, and to impose new and unnecessary costs on the business. This, he argued, was not the function of business managers, but of elected politicians who would ensure that democratically-agreed controls were imposed on business activities. In any case, most claims about social responsibility were likely to be fraudulent (businesses remain primarily profit-making organizations), and dangerous because they pandered to a prevailing anti-business, anti-profit culture which carried long-term risks for business.

The 1990s saw a revival in concerns about corporate social responsibility.[3] Evidence of this is widely available in pronouncements by business leaders and the proliferation of initiatives, many sponsored by governments, that have emerged in recent years. While the causes of the current debate about social responsibility are not well understood, three possible explanations stand out.

First, there has been a concern with business management itself. This is where the corporate social responsibility debate is linked in interesting ways to the corporate governance debate. Management theory in the 1980s and 1990s proposed that internal control and effectiveness in large, global and culturally-diverse businesses required a set of 'corporate values' that could be articulated clearly and would shape the behaviour of employees within the organization.[4,5] Organizational cohesiveness and management control came to be seen as depending on some common set of 'cultural' reference points. This is one way in which the loyalty and commitment of employees could be captured and sustained. These reference points cannot normally appeal to the search for profits alone, but usually connect to a set of broader social or ethical values, many of them culturally specific. Corporate social responsibility is one way in which firms have sought to build a consistent picture of these values as an ideological system that socializes employees to strategic objectives. It may also substitute for rigid approaches to management control by helping to bind employees to these corporate goals. Corporate social responsibility is therefore frequently internally-directed, involving a process of internal transparency and accountability of managers to their workforce (and *vice versa*).

Second, structural and market changes have profoundly influenced the competitive environment in which many firms operate. Greater competition, especially in commoditized markets, has forced businesses to seek new ways of differentiating their products and services (and their 'brand') with the final consumer. The actions that fall under the banner of corporate social responsibility are one way of supporting the construction and defence of 'corporate reputations'

[2] M. Friedman, The social responsibility of business is to increase its profits, *New York Times Magazine*, September 13, 1970.

[3] M. B. E. Clarkson, A stakeholder framework for analyzing and evaluating corporate social performance, *Acad. Manage. Rev.*, 1995, **20** (1), 92–117.

[4] R. M. Kanter, The new managerial work. *Harvard Business Rev.*, 1983, **66** (6), 85–92.

[5] A. W. Pasmore, *Creating Strategic Change*, Wiley, New York, 1994.

and brands. This type of 'signalling' is also oriented at shareholders. As shareholding has become more distributed, both the intensity and variety of public 'signalling' by corporations has increased. Shareholder value is still primarily defined by the growth in share prices, but a number of other means of signalling to shareholders have been developed. These may be seen as a response by management to the need to demonstrate quality and prospects for future share performance, even during periods of weak share performance. Therefore, it can be argued that the greater vulnerability of firms and their managers in more competitive markets has produced a need for new forms of signalling about management performance. Corporate social responsibility and reporting, and other signalling activities associated with them, has been a particularly significant response to this need.

Third, corporate social responsibility may be seen as a response to the changing context of social regulation within which firms operate. The 'statist' model of regulation, in which governments – regional, national or local – impose legally enforceable standards on firms, is being replaced by a model of social regulation that is more interactive and distributed. New information-rich voluntary and market-based regulatory measures are being developed to complement classical systems of 'command and control' regulation. While national environmental policy styles remain highly specific, the capacity of governments to secure the public interest in the environmental field has been reshaped. Increasing voluntarism, 'partnership' between business and government, and a more influential role for non-governmental organizations are all signs of this process of 'ecological modernization'.[6,7]

This new context of social regulation has posed challenges for business, which has sought to develop new capabilities and roles in response. Although firms in many industries have secured greater economic freedoms as a result of the liberalization and deregulation of markets, in many cases this has been matched by a new set of pressures to demonstrate conformance with social norms and expectations, frequently also through a process of re-regulation. Corporate social responsibility can therefore be seen as a way of 'filling the space' that has been opened in the reshaping of environmental governance of firms. Paradoxically, many firms now operate in a more difficult and insecure social environment with a wider range of constituencies to relate to, and expectations to meet. Formal regulation, while often inflexible and procedurally onerous, presents firms (especially large ones) with clear objectives and a simple set of external relationships to manage. In a more fluid and voluntaristic regulatory context these certainties are replaced, and relationships with 'stakeholders' need to be reconfigured.[8]

In short, corporate social responsibility is being shaped by a bundle of needs in many firms: new needs for internal cultural cohesion and management control; new needs to 'signal' about management quality to customers and shareholders; and new needs to engage actively in the new context of social regulation of firms.

[6] A. Weale, *The New Politics of Pollution*, Manchester University Press, Manchester and New York, 1992.

[7] J. Murphy, Ecological modernisation, *Geoforum*, 2000, **31** (1), 1–8.

[8] J. S. Harrison and R. E. Freeman, Stakeholders, social responsibility, and performance: empirical evidence and theoretical perspectives, *Acad. Manage. J.*, 1999, **42** (5), 479–485.

These needs, although related by being the outcomes of linked economic and social changes that have affected business, are each quite distinct. They each face in a different direction, the first towards employees, the second towards customers and shareholders, and the third towards government and other civil society actors.[9] This multiplicity of audiences is commonly aggregated (and confused) through usage of the term 'stakeholders' – giving the impression of a uniform group. In practice, each of these groups invite quite specific management actions. What each of the groups share, however, is the difficulty of establishing clearly the scope of knowledge and accountability that they desire, and consequently the nature of activities that business needs to make. There is no way of defining clearly what responsible behaviour is, or who's view of what constitutes responsible behaviour should be taken into account. There will exist varieties of opinion at any one time, and views will change over time. What is deemed responsible today may appear irresponsible tomorrow. Marginal opinions today can become mainstream opinions in the future. Audiences of corporate responsibility are usually diffuse, not organized and not *active* in the sense of having close and continuous relationships with firms. The contexts within which social norms for business are formed and shaped are therefore ambiguous, open-ended and dynamic.

The mobile and unfocused nature of corporate social responsibility promotes a great variety of responses from different firms, and this apparent variety of responses fuels a counter demand for greater standardization in the *practice* of corporate responsibility. Many firms seek to appeal to groups of stakeholders in a coherent and integrated way through dedicated management functions and though common sets of approaches. Broadly, these have four features:

- values and norms promoted within and by the organization;
- actions consistent with these norms through which responsibility is demonstrated;
- the setting of performance objectives; and
- routines for reporting on actions.

The balance between these features differs between organizations, and also gives rise to differing views about the role of corporate social responsibility. The 'normative' view holds that corporate social responsibility is a response to firms' need to demonstrate social and ethical values that match those of consumers and employees. Corporate social responsibility is seen, therefore, as a means of securing a 'licence to operate'. An alternative 'functional' view sees corporate social responsibility as a way of imposing better management control in business organizations through the introduction of new management and information systems. A simple equation is made between better management control and better corporate performance. Amongst the many influences on this view is the 'balanced scorecard' literature in strategic management.[10] In the environmental

[9] G. Azzone, M. Brophy, G. Noci, R. Welford and C.W. Young, A stakeholders' view of environmental reporting, *Long Range Planning*, 1997, **30** (5), 699–709.

[10] R. Kaplan and D. Norton, The balanced scorecard: measures that drive performance, *Harvard Business Rev.*, 1992, January/February, 71–79.

field this view has been linked to the 'double dividend' arguments of the early 1990s that proposed that firms sought not only to comply with regulations, but actively sought to gain competitive advantages by moving 'beyond compliance'.[11,12]

3 Environmental and Social Reporting

Although a phenomenon of the last five years only, the literature on environment and social reporting by companies is extensive.[13–18] It includes:

Reports (corporate, sectoral, regional, national). Corporate environmental and social reporting is now widespread in larger companies in both manufacturing and service sectors. Rikhardsson estimates that, worldwide, some 7000–10 000 reports are produced annually by firms.[19] They are highly diverse in content and frequently produced by external consultancies.

Surveys and reviews of reports. Both academic researchers and consultancies have conducted 'content' analyses of corporate environmental reports to characterize trends in reporting and to evaluate corporate environmental 'engagement'.[20–23] The question of what is reported and how is the focus of these surveys, frequently leads to rankings of the reporting performance of firms. Implicitly, these rankings are used as a proxy measure for environmental management and performance by firms. This research has plotted the growth in reporting, and has proposed a number of evolutionary trends in the quality of reports (the UNEP Five Stage Model of corporate reporting, for instance). It shows that environmental reports are becoming more like financial reports in structure, but contain highly variable sets of information. The focus of reporting is also seen as shifting from being one-way, passive and unverified, to a two-way dialogue that focuses on impacts and benchmarking.[24]

[11] M. Porter and C. van der Linde, Green and competitive: ending the stalement, *Harvard Business Rev.*, 1995, September/October, 120–134.

[12] N. Walley and B. Whitehead, It's not easy being green, *Harvard Business Rev.*, 1994, May/June, 46–52.

[13] G. Azzone and R. Manzini, Measuring strategic environmental performance, *Business Strategy Environ.*, 1994, **3** (1) (Spring), 1–14.

[14] M. Bartolomeo, *Environmental Performance Indicators in Industry*, 1995, FEEM, Milan.

[15] M. Bennett and P. James, *Sustainable Measures*, 1999, Greenleaf, Sheffield.

[16] D. Ditz and J. Ranganathan, *Measuring up*, World Resources Institute, Washington DC, 1997.

[17] M. J. Epstein, *Measuring Corporate Environmental Performance*, 1996, Irwin, New York.

[18] IRRC, *Corporate Environmental Profiles Directory 1996: Executive Summary*, Investor Responsibility Research Center, Washington DC, 1996.

[19] P. M. Rikhardsson, Information systems for corporate environmental management and accounting and performance measurement, in *Sustainable Measures*, M. Bennett and P. James, Greenleaf, Sheffield, 1999, pp. 132–150.

[20] G. Azzone, G. Noci, R. Manzini, R. Welford and C. W. Young, Defining environmental performance indicators: an integrated framework, *Business Strategy Environ.*, 1996, **5** (1), 69–80.

[21] J. Elkington, N. Kreander and H. Stibbard, A survey of company environmental reporting: the 1997 Third International Benchmark Survey, in *Sustainable Measures*, M. Bennett and P. James, Greenleaf, Sheffield, pp. 330–343.

[22] G. Noci, Environmental reporting in Italy: current practice and future developments, *Business Strategy Environ.*, 2000, **9** (4), 211–223.

[23] A. White and D. M. Zinkl, Raising standardisation, *Environ. Forum*, 1998, January/February, 28–37.

[24] See: www.globalcompact.org

Commentaries on reporting and standard-setting initiatives. Corporate reporting has emerged in a voluntary and *ad hoc* way in response to the specific needs of firms. The diversity that has emerged has consistently been contrasted with the standardization that exists in financial reporting.[25] This is partly due to a lack of regulatory requirements for environmental and social reporting – with only Denmark and the Netherlands implementing mandatory schemes – but also because of positive arguments for retaining voluntarism and diversity in environmental management.[26] From the mid-1990s on, these concerns about standardization led to the spawning of multiple initiatives to set guidelines for corporate reporting (see below for discussion).

Best practice guides. As reporting has become more routine and formalized, there has been a recent proliferation of 'best practice' guides for producing reports and, in particular, performance indicators.[27,28]

4 Environmental Performance Measurement and Reporting

Environmental performance measurement and reporting has been a consistent theme in the literature on corporate reporting. Many reports contain quantitative performance information; surveys and reviews have identified performance measures as being increasingly important through time, and much of the debate about standardization has been over how standard sets of environmental indicators can be derived for firms. James suggests that six distinct frameworks for environmental performance measurement can be identified – production, auditing, ecological, accounting, economic and quality (see Table 1).[29]

James also argues that the diversity of environmental issues, organizational variables (size and management style), national circumstances and individual corporate strategies are likely to mean that performance measurement activities will continue to vary between countries and industries. This prediction has been borne out in the many different reporting approaches and schemes adopted by companies.[30–32]

[25] R. Adams, Are financial and environmental performance related? *Environ. Accounting Auditing Rep.*, 1997, May, 4–7.

[26] P. M. Rikhardsson, Statutory environmental reporting in Denmark: status and challenges, in *Sustainable Measures*, M. Bennett and P. James, Greenleaf, Sheffield, 1999.

[27] BMU and UBA, *A Guide to Corporate Environmental Indicators*, Federal Environment Agency, Bonn, 1997.

[28] D. Wathey and M. O'Reilly, *ISO 14031: a Practical Guide to Developing Environment Performance Indicators for your Business*, The Stationery Office, London, 2000.

[29] P. James, Business Environmental Performance Measurement, *Business Strategy Environ.*, 1994, **3** (2), 59–67.

[30] W. van der Werf, A weighted environmental indicator at Unox: an advance towards sustainable development?, in *Sustainable Measures*, M. Bennett and P. James, Greenleaf, Sheffield, 1999, pp. 246–252.

[31] M. Wright, R. Allen, R. Clift and H. Sas, Measuring corporate environmental performance: the ICI environmental burdens approach, *J. Ind. Ecol.*, 1998, **1** (4), 117–127.

[32] NRTEE, *Measuring Eco-efficiency in Business*, National Round Table on the Environment and the Economy, Ottawa, 1997.

Table 1 Frameworks for environmental performance measurement

Approach	*Orientation*	*Drivers*	*Measurement focus*	*Metrics*
Production	Engineering	Efficiency	Mass/energy balance	Efficiency Resource use
Regulatory	Legal	Compliance	Management systems Non-compliance	Emissions/waste Risk
Ecological	Scientific	Impact	Impact assessment Life cycle assessment	Emissions/waste Impacts Resource use
Accounting	Reporting	Costs Accountability	Liabilities	Emissions/waste Monetary
Economic	Welfare	Internalizing externalities	Environmental valuation	Monetary
Quality	Management	Pollution prevention	Emissions/waste generation	Emissions/waste Monetary

Adapted from James.[29]

5 Standardization Initiatives

The flourishing of corporate reporting saw the emergence of a range of activities seeking to standardize performance measurement and reporting.[33,34] This stress on standardization seems paradoxical given the multiple objectives of performance measurement and reporting, and the evidence of divergent practices in reporting. It is clear that sectors and firms are not alike in their approach to the collection, use and reporting of environmental and social information. Three main sources of pressure for standardization can be identified: the formalization of environmental management; accountancy practice and interests; and 'right to know' advocacy.

Environmental Management

There are several explanations for the stress on harmonization and convergence. The first is the link with environmental management systems. This was prefigured in the EU the voluntary European Eco-Management and Audit Scheme (EMAS) which prescribed an information and reporting system for registered sites. The formalization of environmental management has produced new monitoring and assessment requirements for firms that can be serviced only through structured environmental performance information. A common environmental management system suggests a common environmental information system.

[33] CICA, *Reporting on Environmental Performance*, Canadian Institute of Chartered Accountants, Toronto, 1994.

[34] ACCA, *Guide to Environment and Energy Reporting and Accounting 1997*, The Association of Chartered Certified Accountants, London, 1997.

Accountancy Practice

The logic of an environmental information systems has, in turn, provided another driver towards standardization from the perspective of accounting. Accounting has played a significant role in the development of environmental measures and reports.[35,36] These accounting approaches propose an equivalence between the procedures for developing financial and environmental information for firms. Know-how and conventions in accounting could be transferred to the environmental field, as could the rigour and credibility of financial accounting. Moreover, accountancy could come to occupy new and potentially lucrative territory. Accountants have also identified a need for financial and environmental information to be integrated in the definition of 'eco-efficiency' (*i.e.* environmental–financial ratios) indicators.[37] It is argued that this form of information would improve management decision-making by linking efficiency and impact indicators to cost. Investors and shareholders would then have information directly linked to firms' general report and accounts. As environmental information becomes richer, more reliable, more timely and more accessible, the expectation is that the investment community will increasingly take account of this information in their decisions. This argument rests on the proposition that firms with higher environmental performance may also demonstrate higher financial performance. Advocates of eco-efficiency argue for this link primarily on the grounds of financial savings due to lower resource costs and lower financial risks due to decreased environmental liabilities.

'Right to Know'

Advocates of information as an instrument in environmental improvement have argued strongly for standardization. Here the argument is made that in developed countries traditional standards-based regulations are being complemented by investor and public pressure as a driver of environmental progress in industry. These new 'market-based' pressures demand a much richer set of information flows between the firm and the decision-maker. Similarly, in developing countries elaborate regulatory systems that are enforceable cannot be erected. Here too, progress will be achieved primarily through social and competitive pressures. Disclosure of information to investors and the general public is a key feature of this newly emerging form of governance. Standardized, validated and comparable environmental information about firms is an essential prerequisite for environmental information to become a complement to regulation.[38]

Given the environmental management, accountancy and 'right to know' contexts within which the case for standardization has been made it is not surprising that in this area, as with other aspects of the environmental

[35] R. Gray, D. Owen, *et al.*, *Accounting and Accountability: Changes and Challenges in Corporate Social and Environmental Reporting*, Prentice-Hall, Hemel Hempstead, 1996.

[36] S. Schaltegger, *Corporate Environmental Accounting*, John Wiley, Chichester, 1996.

[37] K. Muller and A. Sturm, *Standardised Eco-efficiency Indicators*, Ellipson AG, Basel, 2000.

[38] T. Tietenberg, 1997, *Information Strategies for Pollution Control*, Eighth Annual Conference EARE, Tilburg, NL, 1997.

Table 2 Environmental performance and reporting standardization initiatives

Initiative	*Scope*	*Participants*
Global Reporting Initiative (GRI)	Global, mainly large companies	UNEP, accountancy organizations, NGOs, academics, business
National Round Table on Economy and Ecology (NRTEE)	Canada, large companies	Business, government
International Organization for Standardization (ISO)	Global, all companies	Business, regulatory, academic
UN-International Standards of Accounting and Reporting (UN-ISAR)	Global, all companies	National experts
World Business Council for Sustainable Development (WBCSD)	Global, large companies	Business
National Academy of Engineering (NAE)	USA	Business, academic

measurement and reporting scene, there are multiple activities. They have included initiatives promoted by government, industry, NGOs in cooperation with industry, standard-setting bodies, and academic institutions. Table 2 presents a summary of the most significant initiatives.

These many standardization initiatives suggest that the demand for common frameworks is widely felt. But the degree of cooperation between firms, regulators and others is noteworthy. It suggests either that performance measurement and reporting is still in a pre-competitive phase, or that clear benefits flow to firms from cooperation. A number of different benefits have been proposed:

A *'learning effect'*. The production of management information and its reporting may be more efficient using standard guidelines. There are costs associated with developing the capabilities to measure and report on environmental and social performance. These costs are likely to be reduced if standard approaches are available.
An *'accounting effect'*. The credibility and accessibility of information, both internally and externally, may be enhanced by cooperation over standards. An accepted, verifiable and auditable set of performance measures, imitating conventions in accountancy, is likely to generate greater legitimacy and trust in the positive 'green claims' that firms may want to make on the basis of environmental reporting.
A *'benchmarking effect'*. The value of information may be enhanced through comparability between firms. Firms may gain greater credit from shareholders, regulators and customers by presenting environmental performance information in comparable form. Benchmarking may also provide greater internal benefits by bringing clarity to the setting of environmental objectives.
A *'coalition effect'*. Self-regulation may pre-empt government regulation. Many governments in Europe have encouraged firms to be more active in their environmental and social reporting, with the implicit threat of regulation if

voluntary initiatives are perceived to be insufficient. Establishing voluntary standards is a way of avoiding or co-opting this pressure.

But despite the strong pressures for standardization, there are also clear tensions. While environmental management-based initiatives have tended to stress the need for a '. . . general, voluntary framework that is flexible enough to be widely used . . .,'[39] accountancy-based initiatives have stressed the need for common frameworks that will be universally applied. These reflect important differences in emphasis, and an implicit conflict over the future environmental management and reporting agenda. Management perspectives place greater weight on management processes,[40] the complex reality of the economic, technological and sectoral contexts of firms, and therefore also voluntarism and appropriateness. Accountancy-based perspectives place more emphasis on the clarity and consistency with which environmental performance information is transmitted by the firm, seeing the firm from the perspective of the balance sheet. Environmental management-based standards (ISO, GRI, WBCSD) have tended to propose broad guidelines that can be linked to evolving internal management systems, whereas accountancy-based standards have been more prescriptive, proposing specific, rather well-defined procedures that can be adopted by all firms.

This divergence in perspective should not mask substantial convergence over end results. Differing perspectives have come to complementary conclusions. This is true for emerging standards over the content of corporate environmental reports, as well as for the sets of performance indicators that have been proposed. In particular, there is now wide agreement over the need for physical performance indicators. Another similarity is that all these schemes avoid single aggregated indicators, preferring instead disaggregated indicator sets. Some alternative indicator sets are summarized in Table 3.

Common indicators include energy and water inputs to production, and global warming, ozone-depleting and solid waste emissions from production. Both GRI and WBCSD also make a distinction between 'core' or 'generally applicable' indicators, and 'supplemental' or 'organization-specific' indicators. This notion of generic and specific indicators is not used in the more externally-oriented WRI, NRTEE and Ellipson schemas. In general, all of these lists are considerably shorter than those proposed when standardization in performance measurement was first discussed, with a range of four to seven generic indicators being proposed.

6 Analysis of Environmental and Financial Performance of Firms

The literature analysing the environmental performance of firms, and the link between environmental and financial performance is broad. Tyteca[41] provides

[39] H.A. Verfaille and R. Bidwell, *Measuring Eco-efficiency: a Guide to Reporting Company Performance*, World Business Council for Sustainable Development, Geneva, 2000.

[40] The ISO 14031.5 standard sets out a system of environmental performance evaluation (EPE) defined as '. . . a management process which can provide an organisation with reliable and verifiable information . . .'.

[41] D. Tyteca, On the measurement of the environmental performance of firms – a literature review and a productive efficiency perspective, *J. Environ. Manage.*, 1996, **46**, 281–308.

Table 3 Environmental performance indicator sets proposed by standardization initiatives

Proposed indicator	*Initiative* GRI* (2000)	WBCSD (1999)	WRI (1997)	Ellipson (2000)
Total energy use	Yes	Yes	Yes	Yes
Total materials use	Yes	Yes	Yes	
Greenhouse gas emissions	Yes	Yes		Yes
Ozone-depleting substances emissions	Yes	Yes		Yes
Total water use		Yes		Yes
Total waste generated	Yes		Yes	Yes
Total pollution emissions			Yes	
Sector-specific indicators	Yes	Yes		

*WRI World Resources Institute

general reviews of indicators used in the analysis of the environmental performance of firms. The following different classes of performance indicator have been used: environmental management; environmental achievements; prevention costs and environmental investment; operating environmental costs; contingent environmental liabilities; physical indicators; and compliance indicators.

Each of these is seen as having advantages and disadvantages from an analytical perspective. Effort-related indicators, including environmental management effort (*i.e.* formal systems implemented) and environmental achievements (*i.e.* environmental prizes), are perceived as simple to define and collect but difficult to relate to environmental performance outcomes. Environmental prevention and operating costs have the same difficulty, and they are also notoriously difficult to determine since the 'environmental' and 'non-environmental' components of a firm's expenditure are difficult to segregate. In principle, all expenditure is in some way related to the environmental performance of a firm. Furthermore, it is unclear whether higher or lower environmental expenditure is an indicator of better environmental performance. A company managing its environmental impacts well will expect lower expenditures through time. These problems also affect the calculation and interpretation of contingent environmental liabilities (future costs associated with taxes, decommissioning of plant and evolution of environmental standards). Physical indicators are seen as better founded and often cheap to generate, but also more diverse, contested and difficult to integrate into decision processes. Finally, compliance indicators are easy to collect, but relate only to a firm's performance relative to legal requirements – beyond that they say nothing.

Tyteca[41] draws out another significant distinction. The first is the question of whether 'simple', aggregated or normalized indicators should be used. Simple indicators express single dimensions of performance in absolute terms (*i.e.* total energy consumed). Aggregated indicators bring together information on a number of dimensions of performance. Many studies of environmental performance have used aggregated indicators.[42–45] Since the mid-1990s, the US Toxics

[42] D. Cormier, M. Magnan and B. Morard, The impact of corporate pollution on market valuation: some empirical evidence, *Ecol. Econ.*, 1993, **8**, 135–155.

Release Inventory (TRI) has spawned a large literature employing aggregate chemicals release information for firms. Tyteca[41] adapted the 'productive efficiency' approach to environmental performance analysis. Productive efficiency indicators are dimensionless expressions of the ratio of undesirable outputs (*i.e.* emissions of pollution), taking into account the efficiency with which inputs are transformed into outputs. Normalized indicators are performance measures that are controlled for some quantity (tonnes of output produced in a year) reflecting the firm's activity.[46]

Explaining Environmental Performance

Studies employing performance indicators have sought both to explain how factors internal (for example, environmental management and technological factors) and external (for example, market and regulatory pressures) to the firm influence environmental performance, and to investigate whether there is a link between firms' environmental and economic performance. Gallez and Tyteca find that firm environmental performance measured using an aggregate 'input-undesirable output' index was explained by investments in abatement technology, total environmental investments in the previous year, and by the age of plant.[47] This latter result is explained by learning effects that become more pronounced through the life of an industrial plant.

A more recent literature has also emerged about the impact of environmental information as a regulatory instrument itself.[48–50] This literature is concerned with testing the effectiveness of more voluntaristic and market-based approaches to environmental regulation. It begins with the proposition that better information is an essential element in reshaping environmental management. By encouraging more information to be made available, firms will come under social and market pressures to improve environmental performance. Using both qualitative (quality of environmental reporting) and quantitative (NO_x and SO_x emissions to air) data, Siniscalco *et al.* found a positive correlation between the quality of environmental information produced by a firm and its environmental performance.[47]

[43] R. W. Haines, Environmental performance indicators: balancing compliance with basic economics, *Total Qual. Environ. Manage.*, 1993, Spring, 367–372.

[44] B. Jaggi and M. Freedman, An examination of the impact of pollution performance on economic and market performance, *J. Business Finance Accounting*, 1992, **19**, 697–713.

[45] J. S. Naimon, Benchmarking and environmental trend indicators, *Total Qual. Environ. Manage.*, 1994, Spring, 269–281.

[46] M. Behmanesh, J. A. Roque and D. Allen, An analysis of normalised measures of pollution prevention, *Pollut. Prevention Rev.*, 1993, Spring, 161–166.

[47] C. Gallez and D. Tyteca, Explaining the environmental performance of firms with indicators, *Ecosystems and Sustainable Development: Advances in Ecological Sciences*, Computational Mechanics Publications, Pensicola, 1997.

[48] P. Lanoi, B. Laplant and M. Roy, *Can Capital Markets Create Incentives for Pollution Control?*, The World Bank, Washington DC, 1997.

[49] M. Khanna and L. Damon, *EPAs Voluntary 33/50 Program: Impact on Toxic Releases and Economic Performance of Firms*, University of Illinois, 1997.

[50] D. Siniscalco, S. Borghini, M. Fantini and F. Ranghieri, *The Response of Companies to Information-based Environmental Policies*, Fondazione Eni Enrico Mattei, Milan, 2000

Another key issue that has stimulated the development of environmental performance indicators has been the question of the relationship between environmental and financial performance. The basic analytical concern has been with whether corporate environmental management is rational in economic terms. Specifically, do market actors (firms, investors and customers) take environmental factors into account in making economic decisions so that there are competitive advantages to be gained through environmental management? The aim has been to test the standard assumption that environmental effort by companies implies a trade-off with financial performance. If this assumption is not borne out empirically, environmental management could be justified in financial terms (either in terms of profitability or in the value of stocks and shares), and the opportunities for integrating environmental objectives into business management would be enhanced. A number of arguments have been proposed for a positive relationship:[11]

- resource savings: more efficient use of resources by a firm will bring cost savings that feed through to higher profitability
- avoidance of environmental liabilities: better environmental management will reduce costs of spills, leaks, accidents and longer-term decontamination and decommissioning costs
- competitive positioning relative to industry standards: 'environmental leaders' are likely to dominate industry-led efforts to set higher environmental standards
- competitive positioning relative to regulatory standards: 'environmental leaders' are likely to face fewer costs in responding to more stringent regulations
- evidence of good management: environmental performance can be seen as a proxy measure of the quality of business management, and therefore also of future profitability (the converse may also be true: poor environmental performance may herald poor profitability)
- greener product innovation: 'greener' companies are more likely to be able to exploit new market opportunities for greener products and services.

Klassen and McLaughlin,[51] Reed[52] and Wagner[53] provide reviews of the literature. Broadly, statistical analysis falls into four categories: event studies; regression analyses; model portfolios; and the addition of environmental variables to existing valuation models. Event studies compare the financial performance of groups of stocks after the announcement of news about a company's environmental performance or regulatory position (either good or poor). Consistently, these studies find that the market penalizes reports of poor performance and rewards reports of good performance.[54] Investors appear to be

[51] R.D. Klassen and C.P. McLaughlin, The impact of environmental management on firm performance, *Manage. Sci.*, 1996, **42** (8), 1199–1214.

[52] D.J. Reed, *Green Shareholder Value, Hype or Hit?*, World Resources Institute, Washington DC, 1998.

[53] M. Wagner, *A Review of Studies Concerning the Empirical Relationship between Environmental and Economic Performance of Firms: What Does the Evidence Tell Us?* Centre for Environmental Strategy, University of Surrey, Guildford, 1999.

[54] J.T. Hamilton, Pollution as news: media and stock market reactions to the toxic release inventory data, *J. Environ. Econ. Manage.*, 1995, **28**, 98–113.

aware of environmental information and to respond to it. Only event studies provide evidence of the causal relationship between environmental and economic performance, indicating that poor (good) environmental performance causes poor (good) economic performance. A link between market reactions to poor performance and subsequent more rapid improvements in environmental performance suggests a feed-back relationship between environmental and economic performance at the firm level.[55] The apparent value placed on corporate environmental information has led to the proliferation of 'environmental ratings' services, including the Dow Jones Group Sustainability Index and the German Corporate Responsibility Rating.

Regression studies explore the statistical correlation between environmental and financial performance.[56] These studies compare indicators of environmental performance with indicators of financial performance for panel data for companies, using a number of different techniques. Most come to a similar conclusion: that there is a small, but statistically significant positive correlation between environmental and financial performance, although this relationship varies by type of industry and form of environmental performance measure used.[57]

Model portfolio studies use the same panel data as regression analyses, but screen out companies with 'poor' environmental performance.[58] The financial performance of the resulting portfolio of firms is then compared with an unscreened sample. The issue being considered here is whether environmental or ethical screens imposed by investors will limit returns, as conventional analyses would suggest. The evidence appears to be that they do not, although not unequivocally. Significantly, some studies show that environmentally screened portfolios can significantly out-perform unscreened portfolios. Last, Feldman *et al.*[59] take conventional models for predicting the value of firms and tests whether environmental and quality variables can improve the explanatory power of the models. The study suggests that environmental and quality variables can add to the power of a model of risk for stocks.

While these statistical studies represent a rich and innovative programme of research, each of them suffers from the same problems as managers and investors – a lack of comprehensive and standardized measures of environmental performance. Statistical studies are generally serendipitous in their choice of data sets, and tend to have been dominated by analyses of US industry where more environmental information is available from official and private sector sources. The situation is now beginning to be improved in the EU with the emergence of environmental ratings agencies, although the derivation of environmental indices remains in many cases opaque. A second problem is that statistical analyses

[55] S. Konar and M. A. Cohen, *Does the Market Value Environmental Performance*? Owen Graduate School of Management, Vanderbilt University, Nashville, 1997.

[56] S. L. Hart and G. Ahuja, Does it pay to be green? An empirical investigation of the relationship between emission reduction and firm performance, *Business Strategy Environ.*, 1996, **5**, 30–37.

[57] S. Johnson, Environmental performance evaluation: prioritising environmental performance objectives, *Corporate Environ. Strategy*, 1996, Autumn, 17–28.

[58] M. A. Cohen, S. A. Fenn and J. Naimon, *Environmental and Financial Performance: Are They Related*?, Owen Graduate School of Management, Vanderbilt University, Nashville, 1995.

[59] S. J. Feldman, P. A. Soyka and P. Ameer, Does improving a firm's environmental management system and environmental performance result in higher stock price?, *J. Investing*, 1997, **6** (4), 87–97.

frequently reflect an ignorance of the conceptual problems associated with the aggregated indicators of environmental performance that they employ. The sometimes cavalier use of indicators undermines the standing and usefulness of these studies' findings. More comprehensive and rigorously developed environmental performance indicators would greatly benefit this school of analysis.

7 Analysis of Environmental Performance in Firms: The MEPI Study

Environmental performance indicators are used in many different ways:

- Business managers use environmental performance indicators as an internal management tool and for external communication.
- Banks and insurers examine the environmental performance of firms to help assess longer-term economic risks.
- Fund managers use environmental criteria to respond to the demand for environmental and ethical concerns to be taken into account in investment decisions.
- Policy makers may evaluate the effectiveness of different policy instruments in improving firms' overall environmental performance.
- Environmental groups compare the environmental profile of firms in order to put political pressure on poor performers.
- Neighbours observe to what extent companies damage their local environment.
- Researchers analyse patterns and trends to improve understanding of the causes of good and poor environmental performance.

In all cases, indicators can provide only partial information that may need to be qualified with information from other sources. Indicators are deliberately simple measures that stand as proxies for complex and often diffuse phenomena. Indicators indicate. Awareness of their specific limitations and biases is an important aspect of their interpretation.

In the next two sections, we report on the results of the Measuring Environmental Performance of Industry (MEPI) study. The study aimed to analyse patterns and dynamics in industrial environmental performance on the basis of publicly available information. To achieve this, a balance had to be found between doing justice to complexity on the one hand, and pragmatism about data availability and quality on the other hand. Environmental performance indicators should aim to compare the comparable. In most cases, this means comparing companies and sites within the same economic sector on an annual basis. For example, benchmarking the energy use of an insurance company with the energy use of a chemical firm may not generate useful results to either.

The MEPI approach distinguished between *variables* and *indicators*. Through literature reviews that characterized the environmental profiles of different industry sectors, and following consultation with representatives from industry, policymaking and financial organizations, indicator sets for each of the sectors studied were generated. These presented identifiable data requirements. Data were collected for this set of variables, providing the information necessary to measure and compare environmental performance. In a second step, indicators were constructed from these variables. In most cases indicators were simple ratios of two variables (*e.g.* water consumption per tonne of paper produced).

A number of standard denominators were used to construct environmental performance indicators:[60]

- Functional unit: a standardized unit of production from a given sector
- Turnover: total sales for a given company (or site)
- Employees: number of personnel employed by the company (or site)
- Value added: total value of sales minus cost of materials
- Profit: untaxed total value of sales minus cost of sales

When comparing indicator sets across sectors, it is clear that many environmental issues are common; for example, energy use, greenhouse gas emissions and water consumption. Other dimensions of performance are specific to production processes; for example, radioactive discharges from nuclear power plants. Therefore, MEPI uses a combination of generic and sector-specific variables. This approach reduces complexity while also providing some flexibility. Details are given in Berkhout *et al.*[61] and Tyteca *et al.*[62]

The main limitations of the MEPI approach were:

1. The approach is based on economic sectors and their different environmental characteristics, but an increasing number of large companies operate in a range of sectors. The sectoral frame and the comparisons that flow from it may therefore need to be treated with caution.
2. The MEPI approach assumes that companies within a sector or sub-sector face similar environmental challenges. This assumption may not hold for sectors with more diverse products and processes (*e.g.* pulp and paper).
3. Environmental data is usually focused on production processes. Currently, it does not provide a means for assessing the environmental performance of goods and services provided by a firm over their entire life cycle.
4. It was not possible to verify data quality or the consistency with which production units were defined (for instance, whether heat and power was integrated with production, or purchased from outside).

8 Summary of Results of the MEPI Study

The Measuring Environmental Performance of Industry (MEPI) study collected environmental and financial data for 270 European companies and 430 production sites in six industrial sectors (electricity, pulp and paper, fertilizers,

[60] X. Olsthoorn, D. Tyteca, M. Wagner and W. Wehrmeyer, Using environmental indicators for business: a literature review and the need for standardisation and aggregation of data, *J. Cleaner Prod.*, in press.

[61] F. Berkhout, J. Hertin, G. Azzone, J. Carlens, M. Drunen, C. Jasch, G. Noci, X. Olsthoorn, D. Tyteca, F. van der Woerd, M. Wagner. W. Wehrmeyer and O. Wolf, *Measuring Environmental Performance in Industry, Final Report to the European Commission*, SPRU, Brighton, 2001.

[62] D. Tyteca, J. Carlens, F. Berkhout, J. Hertin, W. Wehrmeyer and M. Wagner, Corporate environmental performance evaluation: evidence from the MEPI project, *Business Strategy Environ.*, 2002, **11**, 1–13.

textile finishing, book and magazine printing and computer manufacture). These data were cleaned, aggregated (where necessary) and normalized, and analysed using principal component analysis (to identify critical variables), and linear multiple regression (more detail on analytical approaches in Tyteca *et al.*[62] and Wehrmeyer *et al.*[63]).

Analysis confirmed that environmental performance can be adequately reflected by a subset of the variables incorporated in the database. These results have important implications for the statistical analysis carried out. Construction of performance indicators, benchmarking and analysis of explanatory factors was based on those variables that appeared to be both sufficiently available within the dataset and were found to be significantly influential to the environmental performance. The analyses that we were able to perform depended on the sector analysed. Due to lack of data, no further analysis of the computer manufacturing sector was possible. In some of the sectors (pulp and paper, fertilizer) with more heterogeneous processes and products, analysis needed to be sensitive to the problem of comparing apples and pears.

Correlation between Business, Management and Environmental Variables

One aim of the MEPI study was to understand better underlying patterns in business environmental performance. In particular, we were interested in understanding whether there are relationships between aspects of business and management performance and environmental performance (for instance, are more profitable firms higher environmental performers?). Regression analyses were carried out, using the reduced core variable sets only. All regressions were conducted using environmental indicators normalized by 'functional unit (FU)'. Multiple linear regressions were carried out with stepwise entering of dependent variables. A summary of the most significant results is shown in Table 4. The table summarizes many results, a few of which are commented on below.

Rankings and Benchmarking

Firms may be ranked across three core environmental performance indicators, for five MEPI sectors (book and magazine printing, electricity generation, fertilizer production, pulp and paper manufacture and textile finishing). Company rankings are a powerful way of using greater transparency in corporate environmental performance to influence management decisions. However, rankings of performance must be treated with caution. A lower rank does not necessarily indicate poor environmental management. It may be explained by the technological, market or regulatory constraints the firm operates within. For example, its products may require a particularly energy-intensive production process. Besides the company level, results were also obtained at the site level,

[63] W. Wehrmeyer, D. Tyteca, and M. Wagner, *How many (and which) indicators are necessary to compare the environmental performance of companies? a sectoral and statistical answer*, 7th European Roundtable on Cleaner Production, Lund, Sweden, 2001.

Table 4 Correlations between management, business and environmental performance variables in firms surveyed in the MEPI study

	Variable	Pulp and paper	Fertilizer manufacture	Electricity generation	Textile finishing	Printing
Management	EMAS / ISO	No result	Certified firms have: NO_x (+); N(itrogen) (−); Profit (+)	No result	No result	No result
	Invest Rep	Number of employees (+)	No result	Profit (profitable firms report)	No result	No result
Business	Sales	COD (+)	(no regression since Sales are functional unit)	Energy input (−); Municipal waste (+)	CO_2 (+); Water input (−); Total waste (+)	CO_2 (−); Water input (+); Total ink (+); ISO (certified firms larger)
	Profit	COD (+); Total waste (−)	Energy input (−); SO_2 (−); Heavy metals (+)	Total fuel input (−); CO_2 (−)	No result	Total ink (+)
	Number of employees	Energy input (+)	NO_x (−); SO_2 (−); COD (+); N (−); Water input (+)	Renewables (−); Gas input (−), EMAS certification	CO_2 (+); Total waste (+); Energy input (−)	CO_2 (+); Isopropyl alcohol input (−); Total electricity (+)
Environmental	Waste	Profit (−)	Number of employees (+)	No result	Sales (+); Profit (−); Number of employees (+)	Total waste: Profit (−); Hazardous waste: Employees (−); Sales (+)
	Air emissions	Sales (+)	Profit (−)	No result	Number of employees (+)	Employees (CO_2 (+); SO_2(−)
	Water emissions	Sales (+)	Heavy metals (+); Number of employees (−)	Variables were excluded from analysis	Number of employees (+); Profit (+)	No result

Environmental	*Water input*	Number of employees (+)	No result	Variables wre excluded from analysis	Sales (−); Profit (+); Number of employees (+)	Sales (+); EMAS registered (lower input); Employees (−); Profit (+)
	Energy input	Number of employees (+)	Profit (−); No ISO registration	Separate reporting of environmental investment	Sales (+); Profit (−); Number of employees (+)	Fuel: Profit (+); EMAS (higher) Electricity: Sales (+), Employees (−); Profit (+); ISO (lower)

Results obtained from regressions on the MEPI variables at the company level. A '+' in brackets following the variable name means a positive correlation, a '−' a negative regression coefficient. 'No result' signifies that no significant correlations could be found between the variable identified in the left hand column and other variables.

with some particular methodological issues regarding, among other, aggregate indicators. An example showing water use by printing firms is shown in Table 5.

Analysis of Results

Data collected by the MEPI study for a large number of European firms across six industrial sectors provided a basis for better understanding the patterns, dynamics and drivers of environmental performance in industry. From a wide variety of results we highlight four, as follows:

Variability in environmental performance. The data revealed wide variability in the environmental performance of companies operating in the same sector. It was also found that the pattern of variability was not consistent across different dimensions of performance. That is, greater variability was discovered across some performance indicators than others. The range of performance variability for several indicators in book and magazine printers is illustrated in Table 6.

Performance variability may be explained by technological factors (different production processes may be used to produce the same output but with vary different environmental characteristics), the effect of regulation (regulatory pressure may produce greater convergence in environmental performance), and the effect of relative prices (different producers may choose to optimize their facilities differently depending on the price of inputs and pollution control). For example, water consumption across different parts of the paper industry is shown in Figure 1.

Firm size and environmental performance. Another relationship investigated using the MEPI data was between firm size and environmental performance. In general, we would expect larger firms to be better environmental performers because they face stronger regulatory and stakeholder pressures, and because they have greater technical and financial resources to improve performance. Analysis of the data suggested that large companies were not consistently better performers than small companies. There is no clear 'size effect'. In fact, on some dimensions of performance, larger companies appear to perform worse. Figure 2 shows sulfur dioxide emissions from electricity generating companies (note the logarithmic x-axis).

Profits and environmental performance. Many previous studies have investigated the links between profitability of companies and better environmental performance. Our analysis of the MEPI data produced mixed results. For most dimensions of environmental performance no clear link to profitability is evident. Evidence for a positive link is sketchy – companies producing more waste tend to be less profitable in the paper, textile finishing and printing sectors. There are also some perverse results. Paper companies emitting more chemical oxygen demand (COD) to water tend to be more profitable, for instance (see Table 7).

Environmental management and environmental performance. Many companies have introduced environmental management systems as a way of monitoring and

Table 5 Ranking table – printing firms on water use (water input per employee)

Rank	*Firm or business unit*	*m³/employee**
1	Druckerei Rudolph	7.8
2	Georg Kohl GmbH	12.8
3	Monti n.v.	17.4
4	Walter Medien GmbH	19.2
5	Alfred Wall AG	26
6	Brühl Druck & Pressehaus Giessen	26.2
7	Bischof & Klein GmbH	34.2
8	Süddeutscher Zeitungsdienst Aalen	35.3
9	Stark Druck GmbH	41.6
10	Enschede-Van Muysewinkel n.v.	50.3

*Mean value for all available years.

Table 6 Highly variable environmental performance in printing firms (selected indicators)

Performance indicator	*Companies*	*Range of performance across companies (worst performer/ best performer)*
Carbon dioxide per employee	14	69
Sulfur dioxide per employee	16	465
VOCs per employee	15	31
Total waste per employee	39	532
Hazardous waste per employee	25	109
Total ink input per employee	53	418
Organic solvents input per employee	29	8769
Total electricity per employee	46	27
Total water input per employee	44	31

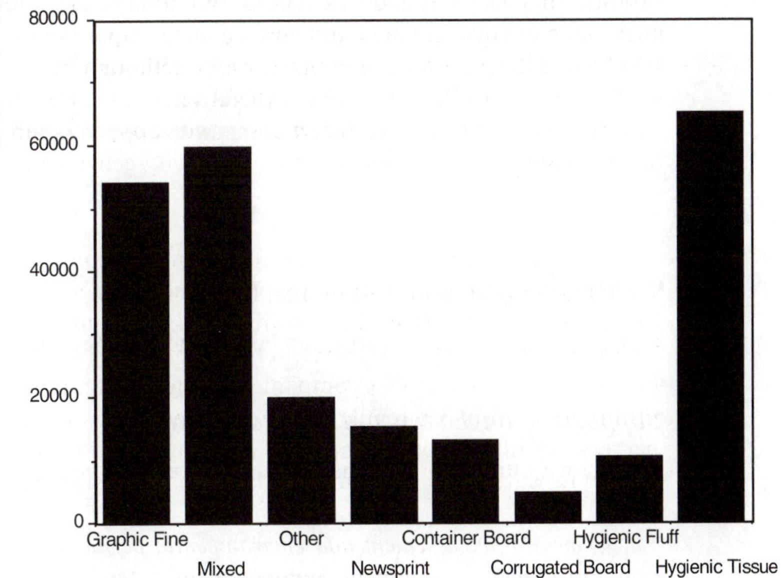

Figure 1 Variable water use in pulp and paper mills (m^3 water $ktonne^{-1}$ of paper, 1996) (Source: Berkhout *et al.*[61])

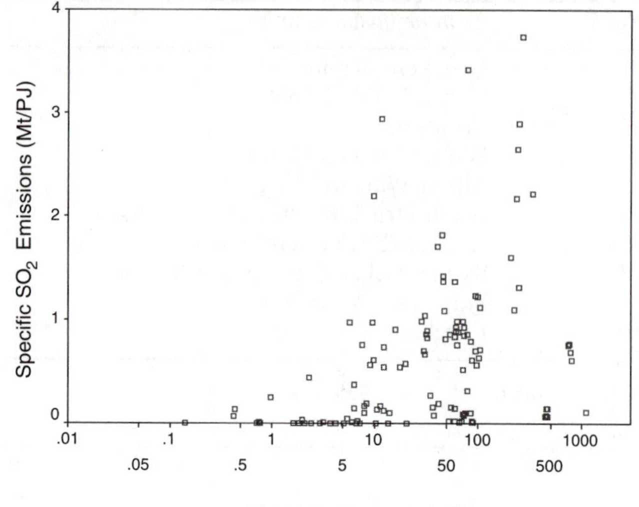

Figure 2 Larger companies are not always cleaner: SO_2 emissions by electricity generators (Source: Berkhout *et al.*[61])

improving environmental performance. We used the MEPI data to investigate whether companies registered/certified with environmental management systems (either ISO 14000 or EMAS) demonstrated significantly better performance.* We found no evidence of a link between environmental management and environmental performance in firms in the electricity, pulp and paper, textile finishing and printing sectors. In the fertilizer sector we found only limited evidence of a relationship: firms with environmental management systems tended to emit more nitrogen oxides to air, and less nitrogen to water. Looking at individual production sites we found evidence of only one positive relationship between environmental management and performance. Paper mills certified with ISO or EMAS had lower COD emissions to water, although this was not a statistically significant result. There was also a negative result. Fossil-fuel based electricity generators with ISO certification perform worse over a number of indicators than those without.

9 Conclusions and Issues for Policy

A number of policy conclusions can be drawn from the results of the MEPI study.

Supporting an Extension of Transparency

There are a number of reasons why transparency is becoming a key principle in environmental policy:

*About 15% of the companies in the MEPI database have ISO or EMAS certified systems.

Table 7 Unclear links between profitability and environmental performance of firms

	Pulp and paper	*Fertilizer manufacture*	*Electricity generation*	*Textile finishing*	*Printing*
Positive correlation with profitability	COD emissions to water	ISO registered Heavy metals emissions to water		Total water used	Total ink used Total water used Total energy input
Negative correlation with profitability	Total waste generated	Energy input SO_2 emissions	Fuel input CO_2 emissions	Total waste generated Energy input	Total waste generated

- Citizens demand the right to know whether companies are behaving responsibly
- Environmental competition between companies (competition on environmental performance, as well as on price and quality) requires a common information, reporting and analytical basis.
- New voluntary and market-based policy instruments are more information-intensive. Markets for environmental services cannot operate without transparency.
- Evaluation of environmental policy impact needs to be based on empirical evidence. Without evidence of benefit commensurate with costs, the legitimacy of policy is undermined.

The MEPI study demonstrated that an information base for conducting integrated analysis of the environmental performance of European industry is becoming available. It has also demonstrated the major weaknesses and gaps that still exist in this information base. The level of performance reporting varies widely between countries, sectors and firms. EU and national-level policy can play a critical role in encouraging and mandating an extension of performance reporting by more firms in more sectors. While governments are beginning to encourage more measurement and reporting, the commitment to these transparency measures remains weak. Widespread benchmarking will enable firms to set targets for improving eco-efficiency, as well as providing incentives for doing so by informing shareholders, customers and regulators about sites' and firms' relative performance.

The process of standardizing environmental data collection, reporting and performance measurement needs to be supported by policy measures. Many voluntary standardization initiatives have produced conceptual frameworks, but few practical tools. Sector-based voluntary and mandatory schemes also need to be considered.

Reducing the Variability of Industrial Environmental Performance

Our analysis confirmed the wide intra-sectoral variability in measured environmental performance (across both regulated and unregulated performance measures) in EU industry. Some of this variability can be explained by technological differences within sectors and some by differing regulatory standards and enforcement. However, these preliminary results suggest that there remain large potentials for improving the eco-efficiency and for reducing environmental impact of European industry.

The results of our study also show that many assumptions about environmental performance need to be revisited. In particular, we have been unable to detect clear links between firm size and environmental performance, and firm location and environmental performance. Small southern European firms seem equally likely, on this evidence, to be good performers, as large northern European firms.

Policy has a role in both widening the scope of environmental performance benchmarking between firms on an EU basis and in using this information for the development of new policy and the implementation of existing policy.

The Uncertain Benefits of Environmental Management Systems

Over the last ten years many companies have adopted environmental management systems, whether registered/certified or not. There has been a wide expectation that these new management approaches would lead to tangible benefits in terms of improved environmental performance. Many companies have argued that 'regulatory relief' should be given to firms with environmental management systems.

The link between environmental management and performance was analysed statistically in the MEPI study. In general, we did not find that those companies with a registered/certified EMS perform significantly better than those without. Indeed, in some cases they appeared to perform worse than those without an EMS. This result suggests that more evidence is needed before environmental arguments are made in favour of regulatory relief for certified firms. It also underscores the need for a better information base for evaluating the impacts of voluntary and market-based environmental policy instruments.

Subject Index